COPING WITH TRAUMA
THEORY, PREVENTION AND TREATMENT

Coping With Trauma

Theory, Prevention and Treatment

Rolf J. Kleber & Danny Brom
In collaboration with Peter B. Defares

CRC Press
Taylor & Francis Group
Boca Raton London New York

CRC Press is an imprint of the
Taylor & Francis Group, an **informa** business

CRC Press
Taylor & Francis Group
6000 Broken Sound Parkway NW, Suite 300
Boca Raton, FL 33487-2742

First issued in hardback 2017

ISBN-13: 978-9-0265-1227-8 (pbk)
ISBN-13: 978-1-1384-6281-6 (hbk)

Cover design: Rob Molthoff

Visit the Taylor & Francis Web site at
http://www.taylorandfrancis.com

and the CRC Press Web site at
http://www.crcpress.com

CONTENTS

Preface

PART I. Introduction and Concepts *1*

1. Traumatic stress *2*
2. General concepts *12*

PART II. Traumatic Events *31*

3. War *32*
4. Disaster *51*
5. Violence *72*
6. Concentration camps *94*
7. Loss *106*

PART III. General Theory *125*

8. Coping with trauma *126*
9. Determinants of trauma and coping *156*

PART IV. Intervention *185*

10. Prevention *186*
11. Behavior therapy *204*
12. Short-term psychodynamic therapy *223*
13. Hypnotherapy *243*
14. The effects of brief psychotherapy *258*

Epilogue *275*

15. Trauma in perspective *276*

References *284*

Index

I Names *306*

II Subjects *313*

PREFACE

In the past ten years the interest in trauma has grown tremendously. Both in the United States and in Europe research efforts have increased and clinical methods have been developed. This book reflects the development in the field.

The book is the product of 10 years of experience in research and clinical practice in the field of traumatic stress. It started with the political hijackings in the Netherlands in the late seventies, when many questions arose about the best method to support the victims and treat eventual disorders. After initial research on theory and treatment, the Dutch Institute for Psychotrauma was established. In this institute further research was combined with the development of outreach programs for people after a variety of severely distressing events and with a continuous treatment facility for people with posttraumatic disorders. In the past years the field of interest has broadened to include, for example, transgenerational aspects of traumatization.

The initial research was conducted at the Dutch Institute for Psychosocial Stress Research. Later the work was supported by the Dutch Institute for Psychotrauma and the Department of Clinical and Health Psychology of the University of Utrecht, The Netherlands.

Over the years, we met many people, who inspired us and whose influence is visible in the book. In particular we like to mention Prof. Mardi J. Horowitz, whose work has been a guideline for us throughout. We thank Prof. dr. J. Bastiaans, Prof. dr. P.E. Boeke, and Prof. dr. J.H. Dijkhuis for their valuable comments on earlier versions of the manuscript.

Many people have contributed to the realization of the book. The many colleagues at the Institute for Psychotrauma, the University of Utrecht and Ezrath Nashim Hospital in Jerusalem have been important sources of support. We thank Ellen Muller for her editorial assistance, and Hedy Kleiweg and Sharon Ben Haim for the translation and correction of the text. The book was made possible by grants from the Dutch Prevention Fund.

Much appreciation is in place for those traumatized people, who taught us, more than books can do, about coping. We hope that this book will help professionals to keep their ears eyes open, even to the experiences no one wants to hear or see.

Utrecht,
The Netherlands.

Jerusalem,
Israel.

PART I

Introduction and Concepts

CHAPTER 1

TRAUMATIC STRESS

Every individual is confronted by drastic events during his or her lifetime. These events are so unpleasant and so shocking that one is barely able to keep going. All of a sudden the security of daily existence has been destroyed. It takes considerable time before an individual is once again able to lead his or her[1] own life without being preoccupied with what has happened.

There are many examples of extreme events: a woman suddenly loses her husband in a traffic accident, a man is beaten up by a gang of youths during his evening stroll. There is often no warning that the event is going to take place. The situation is so normal, so self-evident, yet suddenly something happens that makes the world look completely different: an individual boards a train in the morning and halfway during the journey, he is suddenly held hostage by a group of terrorists. There is no warning and no preparation is possible.

Sometimes extreme events concern coincidences that simply occur every now and then. A truck driver with years of experience kills three people in successive car accidents within a few months, without any of the deaths having been his fault. A woman sees her child climbing a tree in the yard and knocks on the window in warning. The child is startled and falls from the tree. These dramatic events are not completely avoidable. They simply occur.

When the events have passed, it does not mean that the experience is over for those involved. The individuals have to deal with aftereffects for a long period of time. The event surfaces in their dreams and in suddenly recurring memories. They feel confused. All sorts of doubts arise about

[1] Traumatic stress does not distinguish between men and women. For stylistic reasons we do not consistently use 'his or her' and 'she or he' throughout the book.

things that were so obviously secure in the past. They are depressed but simultaneously restless as well. It is difficult to accept all the changes that the drastic event brings with it. The victims are angry about what has happened to them and feel they have been abandoned. They would prefer to pretend the event did not occur, but they are continually confronted with the irrevocability of it. They are afraid: afraid to be alone, afraid that the disaster will recur, afraid of everything that makes the memory surface. Thus they avoid all that concerns the event.

All these consequences have their effect on the lives of those involved. After a while, most people manage to cope with the shock, even if the memory does not disappear. Some, however, have such difficulty with the painful experience that they seek outside help.

This book deals with extreme life events, which are situations that consist of experiencing or witnessing serious human suffering, in which there is no way to prevent them, and in which confrontation with death and violence plays an important role. Examples are natural disasters, crimes of violence, acts of war, accidents and the sudden loss of a loved one.

The general goals of this book are:

1. A systematic analysis of research and theories regarding the consequences of and the coping with serious life events, based on psychological, sociological and psychiatric literature;
2. An analysis of intervention strategies and psychotherapeutic treatment methods for shocking events (also referred to as traumatic events). The aim is to arrive at an integrative view of coping and mental health care regarding traumatic stress.

First, we present a phenomenological description of a traumatic experience. Unless noted otherwise, all examples in this chapter are based on clients from our study of treatment methods of posttraumatic stress disorders (Brom, Kleber & Defares, 1989).

1.1 What is a traumatic experience?

Mr. D is a 38-year-old salesman. Some time ago he returned home at night from a family gathering held outside of the city. His wife sat next to him in the car. Sitting in the back were his two children and his father, who was staying with him for a few days. The road was barely lit and there was a light drizzle. On the way home he had an argument with his father, who felt he was driving too fast. His wife quieted them down. D was a good driver and, after all, he had not had anything to drink.

Suddenly, while going around a curve, a large truck appeared on D's side of the road. His wife screamed. He attempted to avoid the truck, which was no longer able to reach its own side of the road. The truck hit the small car, which rolled across the road into the soft shoulder and landed in a pasture.

Later D could not remember just how long he had lain unconscious in the car, but suddenly he heard the voices of strangers. He looked around in a daze. His wife was gone and behind him he heard moaning. He was removed from the car by first aid attendants and carried to an ambulance. Out of the corner of his eye he could see his father lying in a strange position in the back of the car. With the sirens wailing, he was taken to a hospital. It was there that he heard his wife had been thrown from the car and had died instantly. His father died after a few hours in the hospital. Both of his children were critically injured. D himself only had a few minor injuries.

The event described above is shocking. It is extraordinary and unpredictable. Without any preparation, an individual is overwhelmed by such a dramatic event. Loved ones pass away. From one moment to the next, a life changes completely. A person's self-image is also disrupted. The man is no longer a husband and a son, but a widower. His own identity changes.

A shock such as the one above can be described with numerous terms. Our language knows many adjectives concerning extremely serious events: horrible, intense, extreme, drastic, etc. Freud (1920) referred to excessive stimulation, which penetrated an individual's protective shield. The stress literature refers to intensity (Baum & Singer, 1981) and stressful life events (Holmes & Rahe, 1967). What does the intensity or overstimulation consist of?

Three aspects are characteristic for an individual in an extreme situation:

1. powerlessness;
2. an acute disruption of one's existence;
3. extreme discomfort.

These three aspects offer a phenomenological description of an individual in an extreme situation. Below we will examine them more closely.

Powerlessness

In an extreme situation, an individual barely has any influence upon the occurrence and development of the event. Acts of war and natural disasters are so overwhelming that the individual is completely helpless. The same applies to accidents and sudden deaths, as in the following example.

A man bicycles home from work. In the distance he sees his son on the sidewalk in front of his home. He waves and the child notices him. Suddenly, without looking, the little boy runs into the street to greet his father. A passing car cannot stop in time. Brakes screech, and the impact sends the child flying through the air. The boy dies almost instantly.

The father feels completely helpless in this situation. He appraises the circumstances as beyond any personal control and realizes that he lacks any grasp of the event, which could not have been prevented. The event is an unavoidable, unpreventable experience that completely overwhelms him.

In 'Jenseits des Lustprinzips' (1920), Freud posed: 'The essence of a traumatic situation is an experience of helplessness that is brought about either externally or internally'. The feeling of not being able to exercise any influence (control) upon a situation and the resulting helplessness are important topics in modern psychology. This relevance has already been pointed out in older literature concerning extreme events. In this book we will often refer to the concept of control.

Acute disruption

A serious life event throws the individual, often from one moment to the next, into a completely different situation, which crudely disrupts the course of daily existence. We already mentioned the example of the passenger who innocently gets onto a train and is unexpectedly confronted with a train hijacking, causing him to spend the following week or two living with continual fears and threats. Having been taken hostage hurls him into a situation of complete confusion, insecurity and desperation (Kho-So, 1977). In an analysis of the post concentration camp syndrome, Bastiaans (1974) pointed to this same process of being cut off from the previously secure environment.

Another example of acute disruption is the sudden death of a loved one. A man and woman return home after a busy evening of shopping. The man sticks the key into the front door and without making a sound, suddenly slumps to the ground and dies almost immediately as the result of an acute heart attack. The event drastically and permanently changes the life of the woman. Nothing seems as it was for her before.

By acute disruption we mean sudden, extremely drastic changes in daily life. In one fell swoop, the self-evident continuity of existence (and thus also the conceivability of one's own future) is destroyed. The existing certainties of life have disappeared. The world of someone who has been struck by a shocking or traumatic event suddenly looks different. The images he holds of himself and his environment no longer adequately fit the new situation.

Extreme discomfort

In this study, the emphasis is upon extremely distressing events. It may be trivial to point out, but the two characteristics detailed above do not sufficiently describe the situations we will be treating. Being passionately in love or winning a lottery are also, after all, examples of experiences of powerlessness that can throw one's existence into turmoil.

In stress research from the sixties and the seventies, such as in the 'stressful life events' approach, it was initially assumed that the adaptation required by traumatic events influenced psychological and physical functioning regardless of the desirability of the events. A wedding would thus trigger as many stress reactions as a divorce. Since that time, various studies have examined the separate effects of pleasant and unpleasant changes in one's personal life (Kleber, 1982). It was found that only the amount of *unpleasant* events was significantly related with psychological problems and illness. The number of positive life events, no matter how fundamental, were barely related to malfunctioning of any kind. Hence it was not a matter of life changes as such, but specifically of unpleasant or negative changes.

Utilizing the constructs of helplessness, disruption and discomfort, we have globally characterized a traumatic event. These three elements are the result of an interaction between the demands of the environment on the one hand and the skills, expectations and characteristics of the individual on the other hand. Some circumstances lead to such discrepancies more readily than do others; one individual more readily experiences an event as traumatic than another. We will often return to this interaction.

1.2 The design of the book

In this book we present a general and systematic perspective on the responses to, and the coping with, traumatic events. It aims at an integration of theoretical models and research findings derived from the scientific literature.

Our perspective is psychological. Social and societal consequences, as well as physical effects such as injuries, will be discussed only in relation to the psychological process of coping. The same goes for the interaction between physiological processes and the reactions to traumatic events. Furthermore, we focus exclusively on events in adulthood. Traumatization in childhood through abuse, early losses and other extreme situations is beyond the scope of this work (see Eth & Pynoos, 1985).

A general chapter will preface our analysis of findings from scientific literature. This chapter will discuss the various concepts that have been used over the years to understand disorders resulting from serious life events. The aim is to clarify the history of scientific thinking regarding this subject matter.

The four following questions are central to this book. The first question is: *How does a serious life event affect the individual?*

It is necessary to deal with this question extensively. The research on the psychological consequences of serious life events makes a scattered impression. Although the number of publications is numerous, there are only few comprehensive and systematic studies of the psychological effects of traumatic experiences.

Moreover, the research field makes a fragmented impression because the publications are written in different 'languages', so that the similarity is minimal. Publications from the forties refer to neurasthenia, fixation and neurosis. Authors writing about Vietnam veterans discuss phobias, stress response syndrome and drug abuse. In research about natural disasters, sociologists employ other concepts and often arrive at findings that are contrary to those of psychologists or psychiatrists. There are considerable differences between the methods, concepts and viewpoints of the different groups of researchers (who often do not cite one another).

Drawing conclusions is made more difficult by the disordered nature of research in the field. We will therefore describe the effects of extreme life events separately according to category. Additional support for our approach is the lack of empirical research into the differences and similarities in the effects of the various types of traumatic events.

Although the existing research may be incomplete and confusing, an analysis is possible. Studying a large variety of publications, may lead to worthwhile conclusions, if only because it is unlikely that all studies are troubled by the same problems. It is therefore worthwhile to study what research conducted thus far has provided in terms of knowledge about the effects of traumatic events. A review of this material is presented in chapters 3 through 7. Each chapter analyzes the main body of evidence, and examines some specific theoretical issues.

We will limit ourselves to those events that are dramatic for nearly everyone. The following categories of events are discussed:

1. extreme situations as the result of acts of war: experiences on the front and other forms of 'combat stress';
2. collective disasters and accidents, such as natural disasters (earthquakes and floods) and 'man-made disasters' (mine disaster and shipwreck);
3. crimes of violence in times of peace, such as rape, hijacking, robbery, assault and battery;
4. long-term extreme situations as the result of war, particularly being imprisoned in a concentration camp;
5. the loss of a loved one: the death of one's spouse or child.

The second question is: *How does an individual cope with a traumatic experience?* This question is based upon two important notions.

First, an analysis of only effects is too limited. Stress research (Kleber, 1982; Lazarus, 1981) has made it evident that the psychological symptoms after an event (such as the fear of visiting the scene of the event), are both a reaction to a shock and an effort to adjust to the changed situation. Both these aspects can be distinguished in traumatic stress symptoms, although sometimes the reaction to the shock (fear as an effect) will predominate, and at other times it will be more the attempt to adjust to the shock (avoidance of the location as a coping mechanism).

The second reason for studying the general coping pattern following a serious life event, is the realization that the majority of studies in the field of traumatic stress research are based on patients; on those who seek help. Thus, most concepts and theories of coping are developed based upon the study of disorders. To understand the reactions to extreme life events, research should not limit its focus to just patients. The perspective should shift from the specific study of disorders to the general coping process after a traumatic event.

Chapter 8 reviews the process of coping. Central in this chapter is the thought that a serious life event does not fit with the expectations and images the person has about himself and the world. He cannot integrate the experience, making his existing frame of meaning hopelessly insufficient. He is assailed by questions such as 'why me?' and 'what is the meaning of it?' Coping consists of a person looking for meaning, and his effort to regain a sense of mastery in life.

This emphasis on coping leads to a much closer look at the entire period of time of the reactions to a shock. A development takes place within the individual. Effects and coping behaviors are different one month after the event than they are one year after the event. This is why the period of time involved has been an essential starting point in the literature, as well as a central issue in the question of the long-term consequences involved.

The chapter about coping is particularly inspired by cognitive directions in stress research. The individual is considered an organism that can process and interpret information. A person does not passively experience an event but is constantly involved in a process of attaching meaning to the event in order to regain a grasp upon his existence. This cognitive perspective on adaptation to stress relies heavily upon psychoanalytical concepts, not in the sense that terms such as ego and libido are adopted but in the emphasis on defensive manoeuvres. Psychoanalytic theory will therefore be mentioned at various points in this work, in the historical sections such as chapter two as well as in the treatment of the theory of Horowitz, whose work comprises a central role in this study.

The third question in this book is: *Which factors influence the consequences of traumatic events and the process of coping with them?*

Considerable differences in reactions and coping behaviors exist. Although two people undergo circumstances that may seem identical, one individual may go to pieces while the other may not. When both the healthy and disturbed ways of coping receive research attention, then naturally the question arises as to the sources of the difference between the two. Since most studies focused solely upon patients and thus only upon serious disorders, the attention paid to these individual differences has been minimal in research.

Coping and traumatic stress reactions are influenced by various determinants that we have divided into three groups: individual characteristics (such as biographical features and childhood experiences that lead to personality traits), social factors (such as the support from the immediate environment) and characteristics of the situation in which the event took place (expectedness, for example). These determinants will be discussed in chapter 9.

Our starting point is the multiple determination of individual behavior (Duyker, 1972). Aspects of the event and its context, the individual and the social environment mutually determine the severity and extent of the consequences and of the symptoms of coping.

The fourth part of the book deals with the question: *How can people be helped to cope after traumatic stress?* This part focuses on methods of intervention after traumatic events and methods of treatment of disorders in coping with traumatic stress.

As noted, a considerable number of individuals experience problems in coping with traumatic events. The images continue to intrude and the fears seem to increase rather than decrease. Physical complaints often include hyperventilation, headaches and other forms of pain. The individual's entire concentration seems to be focused on the event. These disorders have drawn attention in recent years under the term of posttraumatic stress disorders.

Chapter 10 deals with issues which are helpful in designing preventive outreach programs for victims of traumatic stress. This a relatively new area of attention in traumatic stress studies. After a summary of our theoretical approach to the process of coping, we analyze the various aspects of psychological assistance after such events. We discuss the goals of psychological interventions and different components of intervention programs. Finally, we describe a specific program of psychological assistance that has been developed and applied with employees of organizations who have become victims of violence, such as in bank robberies and hijackings. This part of the chapter is based on the practice of preventive counselling programs which are developed in The Netherlands.

In chapters 11 through 14, we focus on disorders in coping with traumatic events. The main question is to what extent these disorders and the symptoms linked to them can be cured by psychotherapeutic treatment. Since every form of psychotherapy is based upon theoretical assumptions, we discuss the theoretical basis of treatments. Although the link between the theory and the practice of psychotherapy is not always clear and sound, psychotherapy without theory is unthinkable. Theory provides the basis for systematic clinical work.

The therapist who is confronted with clients suffering from posttraumatic stress disorders needs a theoretical framework. It is not sufficient to merely have knowledge of the factors that generally influence the coping process and the development of symptoms. The therapist must have a theoretical model which gives him the opportunity to search for the origins of the patient's disorder and to utilize this understanding, explicitly or not, in treatment.

The three treatment methods that will be discussed stem from different backgrounds. In three separate chapters, we will delve into theory as well as practice. In chapter 11, the reaction to traumatic events is described from a learning theoretical viewpoint. This chapter reviews the classical conditioning theory as well as the contribution of the cognitive behavioral therapy. A treatment method that is linked to both will be presented in the remainder of the chapter.

Undoubtedly, the most elaborate theory about the effects of traumatic experiences derives from psychoanalysis. Over the past 15 years M.J. Horowitz has utilized several of the psychoanalytic concepts involved in traumatic experiences to develop a cognitive theory. This theory is supported by ideas from the stress literature and the literature concerning information processing. Horowitz's treatment method will be discussed in chapter 12.

Hypnotherapeutic techniques have recently received increasing attention, after many years of practice by a very small group of therapists (Van der Hart, 1991). Hypnosis can, in fact, be a component of various forms of psychotherapy. In chapter 13, we describe some theories and the way in which hypnotherapy was applied in our work. In this chapter as well as in chapter 11 a specific protocol for treating patients with disorders in coping with traumatic stress is included.

In chapter 14 we describe a study of the effects of treatment of disorders in coping with extreme experiences. This comparative outcome research involved three forms of psychotherapy: trauma desensitization, hypnotherapy and psychodynamic therapy. All of the clients who participated in this study had experienced a traumatic event an average of two years earlier. The disorder was triggered by the sudden loss of a member of the immediate family or by experiencing a violent crime or a traffic accident. Important questions examined are: Which changes take place as the result of the therapies that were utilized? Do these effects differ from the changes that take place in a control group? Which clients have benefited from a therapy and which have not? Which indications or contra-indications can be made for assigning someone to one or the other form of therapy?

In this book we provide a perspective on trauma that is characterized by an emphasis on:

- the normal development in time of the consequences of a serious life event;
- commonalities in response patterns across different traumatic events;
- the general process of coping, and the occurrence of disturbances in this process;
- the individual process of giving meaning to the event;
- the multiple causes of post-trauma reactions and coping.

In chapter 15 we review this general theoretical framework on traumatic stress and connect it with insights about the nature of psychotherapy. Both coping with traumatic events and undergoing psychotherapy are activities in which people attempt to assign a new meaning to past experiences.

CHAPTER 2

GENERAL CONCEPTS

Over the years, various concepts such as trauma, traumatic neurosis and posttraumatic stress disorder have been developed to describe the consequences of extreme experiences. These general concepts were both the result of, and impetus for, extensive theoretical developments that were strongly influenced first by psychoanalysis, and more recently by cognitive psychology and stress research. The assumptions concerning these concepts and related issues have played an important role in the study of traumatic events. This influence is one that is still valid today.

It is useful to analyze these concepts before discussing the various types of traumatic events. This will clarify the interrelation of symptoms, while it may also highlight some important issues in this area of research. In this chapter, we discuss and compare general concepts which appear in the literature in order to evaluate them. More specific concepts will be addressed in the next chapters. The historical perspective of this chapter can clarify the various ideas, issues and their development. We will discuss: Freud and his predecessors, the World War II period, and modern developments.

2.1 Freud and his predecessors

In the 1880s, physicians developed an interest in the effects of traumatic events. The German physician Eulenburg, according to Van der Hart (1991), introduced the term psychic trauma in 1878, as a designation for the reaction of outcry and fear after extreme shock. Attention was especially given to the so-called 'Erichsen's disease', which occurred following railway collisions or the abrupt stopping of a train (Keiser, 1968). In 1866, Erichsen described this condition as 'symptoms following (train) accidents which may assume the form of traumatic hysteria, neurasthenia, hypochondriasis or melancholia' (Keiser, 1968, p. 15). He attributed this condition, also labelled as 'railway spine', to a concussion of the spine resulting from the accident.

Physicians in those days vehemently debated the causes of emotional responses to an accident. Some of them assumed the causes were of a purely physical nature. Oppenheim, one of the most prominent physicians of that time, attributed the emotional responses to molecular changes resulting from electrical processes in the central nervous system. But the British surgeon Page, in complete contrast to Erichsen and Oppenheim, drew a distinction between physical injuries often due to hemorrhages within the spinal cord and symptoms, in which psychological events were of primary importance. In 1883, he introduced the concept of 'nervous shock' which was essentially psychological in origin (Trimble, 1981).

The term traumatic neurosis (in German known as Schreck Neurose) emerged from the literature of the 1880s. Oppenheim is thought to have been the first to use this term (Culpan & Taylor, 1973). In his study of the neuropsychiatric effects of injuries, he distinguished four conditions: hysteria, neurasthenia, organic syndromes and traumatic neurosis. These conditions resulted from the effects of injuries, especially those to the head. According to Oppenheim, the causes were physically determined (Keiser, 1968).

Other physicians, however, assumed that emotional reactions to accidents and injuries were of a primarily psychological nature, resulting from anxiety, tensions and suggestions. The most prominent proponent of this psychogenic explanation was the French psychiatrist Charcot. Like most physicians who took part in the debate, he was fascinated by a striking response: namely that of conversion symptoms or hysteria. Charcot distinguished a special form of hysteria – traumatic hysteria – which resulted from a shocking event (Bally, 1969).

It is not easy to determine exactly what was meant at that time by the term traumatic neurosis. It is clear, however, that the term was used in an organic context until World War I, following Oppenheim's theory concerning molecular change (Keiser, 1968). The confrontation with the symptoms of soldiers put an end – at least partially – to this purely neurological explanation, as we will see later on.

Pierre Janet

Charcot encouraged Pierre Janet (1859-1947) to come to Paris to continue his investigations of hysterical patients. In many meticulous clinical observations this French physician investigated the mental processes involved in hysteria and other forms of psychopathology. Janet was the first to conceptualize the influence of traumatic events upon psychological functioning.

His key concept was dissociation: the splitting off of thoughts, actions and feelings from conscious awareness and voluntary control (Janet, 1889). A person responds to an overwhelming experience with vehement emotions, such as intense anxiety and anger. If the person fails to master these intense emotions, he may react with dissociation. The memories of the event cannot be integrated into his memory system. Subsequently, traces of the intensely arousing experience may subconsciously affect behavior and result in various disturbances, such as amnesia, narrowing of consciousness, obsessional preoccupations and somatic symptoms. The impressions of the traumatic event have bypassed consciousness and continue to plague the individual as unrecognized and unintegrated memories: 'the person is unable to make the recital which we call narrative memory, and yet he remains confronted by (the) difficult situation' (Janet, 1925, p. 1). To overcome the traumatic reminiscences and the feelings of helplessness the experience has to be 'liquidated', that means: transformed into a personal narrative.

Although Janet continued his work, his studies were nearly completely forgotten till only a few years ago. He did not establish a school of psychiatry personally, and his ideas did not exert a large influence upon the course of scientific thought. However, recently there has been a revival of interest in the issues he was concerned with, such as dissociation, hypnosis, and the storage of traumatic experiences in memory (Van der Kolk & Van der Hart, 1989).

Sigmund Freud

In 1885, Freud attended Charcot's lectures for a while and became familiar with Janet's early work as well. Charcot's work deeply impressed the Austrian psychiatrist, who had only just decided to discontinue his neurological research. After his return to Vienna he came in contact with the general practitioner Breuer. Together they worked from Breuer's notes regarding the treatment of Anna O., who developed emotional problems after the death of her father.

Breuer and Freud discovered that the etiology Charcot had formulated for a specific form of hysteria applied to all forms of hysteria. The hysterical symptoms were related to an emotional event that had taken place at some time during the life of the person. The symptoms were the remnants, the 'memory symbols', of that traumatic experience. The experience that had evoked the original emotions, that were subsequently converted into physical symptoms, was referred to by Breuer and Freud as the psychological trauma. This trauma – thus the experience of an event – determined both the occurrence of hysteria and the nature of the symptoms (Nuttin, 1968).

According to Breuer and Freud, the patient had suppressed strong emotions during the traumatic experience. This emotion could not have been expressed in words or actions. Furthermore, the event itself had been forgotten by the patient in day to day life. Relief from the hysterical symptoms, according to Freud and Breuer, could take place once the patient could discuss the concomitant trauma with an emotional release. The affect evoked by the psychological trauma still needed to be vented.

The collaborative work of Freud and Breuer was primarily concerned with recent memories, such as the death of the father of Anna O. Intrigued by the free association technique and the trauma theory, Freud continued on this track and encouraged his patients to delve further back into their memory. As of the 1890s, the sexual experiences of patients during childhood increasingly became the central focus of his work and he devoted less and less attention to their recent experiences. In 'Zur Aetiologie der Hysterie' (1896), Freud indicated how, during the treatment of several hysterical patients, he discovered they had been sexually abused in early childhood by adults, older brothers or sisters. These traumata had been experienced prior to adolescence and constituted an actual stimulation of the genitals. Hysterical symptoms were supposed to stem from actual sexual events. This is the so-called seduction theory with regard to hysteria (Buelens, 1971).

Freud gradually began to doubt this trauma theory. He began to suspect that the patient's story about seduction and abuse during childhood was the product of sexual desires and fantasies in that period. Around the turn of the century, Freud gradually began to postulate his theory about infantile sexuality. Whether this theoretical change was also influenced by other reasons such as Freud's own unease with the trauma perspective or a societal rejection of a viewpoint emphasizing sexual abuse is still a matter of debate. An actual traumatic experience was no longer the point of departure; instead the focus was upon the sexual and aggressive wishes of the child. References to traumatic events are therefore rare in Freud's work after 1900 for a substantial length of time.

In Freud's work stemming from the period between 1890 and 1900, 'psychological trauma' refers to an unpleasant event that evokes anxiety, pain, fright and shame. The emotions about this event must be suppressed. The term traumatic neurosis is used incidentally. In 1893, Freud and Breuer postulated that a traumatic neurosis is preceded by a single shocking event, while hysteria is triggered by a number of partial traumata, each having to do with one case of suffering (this distinction reappears in modern psychoanalytical literature on 'shock trauma' and 'strain trauma'; see Furst, 1967).

Between 1914 and 1918, World War I raged on in Europe. Confrontation with the emotional problems of soldiers resulted in renewed interest in the consequences of extreme events, among academic psychiatrists as well as among psychoanalysts, who were still relatively rare at that time.

The previously mentioned two types of explanation for symptoms – the organic and the psychogenic explanation – once again became the focus of attention. At the beginning of the war, the generally accepted view was that emotional problems resulted from physical causes. Mott (Keiser, 1968) substituted the term traumatic neurosis with 'shell shock' because he believed the issue was physical brain damage resulting from a displacement of air, an overdose of carbon monoxide and flying shrapnel. This explanation was gradually abandoned during the war. Soldiers who had not experienced shellfire were also susceptible to this 'shell shock', while patients would recover when they were removed from the trenches. Gradually the psychogenic explanation received a dominant position in academic psychiatry, which was barely influenced by psychoanalysis in the first decades of this century.

The end of World War I clearly affected Freud's thinking. Experiences with war victims were partially at the root of the structural model of Es, Ich and Ueber-Ich (Id, Ego and Super-ego), which was formulated at this time, and of the postulation of the death wish.

In 'Jenseits des Lustprinzips' (1920), Freud formulates the concept of 'Reizschutz' (stimulus barrier) and considers all those stimuli that are strong enough to break through this protective shield as 'traumatic'. The concept of a threshold at which an individual breaks down as a result of over-stimulation regularly appears in later literature.

In 'Hemmung, Symptom und Angst' (1926), Freud returns to the term 'traumatic' and relates it to the origin of fear. A traumatic situation refers to experiences that are unsatisfactory and involve helplessness, in which stimulation reaches such an unpleasant level that it can no longer be controlled. The related notion of loss of control, for that matter, often appears in modern literature.

Freud was especially intrigued by the compulsive repetition that is so characteristic of the effects of traumatic events. Time and time again, victims think back to the moment of the accident or disaster, although their expressed desire is to forget the trauma. It was specifically this compulsion that resulted in Freud formulating the 'death wish,' although at the same time he often emphasized the contribution of this repetition to the coping process (see chapter 8).

Only sporadically does the term 'traumatic' appear in Freud's later work. It often indicates that Freud did not know (yet) how to deal with the nature and effects of traumatic experiences. Sometimes he considered a traumatic neurosis as merely a form of neurosis but sometimes he also saw it as a separate category alongside all other neuroses.

2.2 The World War II period

Despite the renewed attention during and immediately after World War I, the interest in extreme events rapidly vanished again. Not until the forties, during World War II, were traumatic neuroses and similar concepts once again addressed. A large wave of publications appeared about emotional problems resulting from the war. By this time, however, psychoanalysis had clearly influenced psychiatry.

Traumatic neurosis

Traumatic neurosis is a term often used in the psychiatric literature of the World War II period. Kardiner (1941) isolated this category from all other neuroses. The core of a traumatic neurosis was a 'physioneurosis' (unlike a transference neurosis, this referred to an actual threat). After the traumatic experience, the control of the ego over the body was impaired. The futile attempt of the ego, with its inadequate resources, to overpower the dangerous stimuli, resulted in aggression and irritation. Kardiner distinguished these acute disorders of control over the body from already existing neurotic problems experienced by the patient, problems that surfaced when the traumatic neurosis reached the chronic phase.

Other authors (Fenichel, 1945; Grinker & Spiegel, 1945), however, contended that traumatic neurosis was not a unique psychological disease, but that it consisted of neurotic reactions similar in cause and effect to all other neuroses, and distinguishable only by the gravity, acuteness and nature of the events. Grinker and Spiegel emphasized that all individuals are limited in the skills necessary to adjust to an extreme situation, but that each reaction to a trauma is contingent upon the individual's personal characteristics and experiences.

In some cases a distinction was made between acute and chronic traumatic neurosis. The former concerned the immediate neurotic reactions to overwhelming fright, an accident or an event of war. Although specific symptoms could differ, the symptoms of chronic traumatic neurosis described by Fenichel (1945) were dominant:

1. decrease in or blocking of various ego functions (e.g. reduced interest in sexuality, more dependent behavior);
2. spells of uncontrollable emotions, particularly of anxiety and rage;
3. sleep disturbances, dreams in which the event is experienced again, and repetitions of the traumatic event in the form of fantasies, thoughts and feelings;
4. psychoneurotic complications.

After World War II, the distinction between acute and chronic traumatic neurosis was used for obvious reasons – after all, the chronic disorders only appeared after the war – but this dichotomy was not generally accepted.

In publications concerning traumatic neurosis, the role of neurotic predispositions is often questioned. The extent to which conflicts originating from childhood and other determinants deemed important in psychiatry, contributed to the existence of traumatic neurosis was vehemently debated, especially during the forties (chapter 3). Hysteria could be related to a sexually charged conflict dating from the early years of the patient's life. Against the background of the atrocities of war, the significance of this factor appeared to be less apparent, just as the distinctions between normal and abnormal, healthy and unhealthy became unclear in such a situation. The external as well as the recent nature of the (possible) causes, placed traumatic neurosis outside of traditional psychopathology.

The position of the concept of traumatic neurosis in the psychiatric and psychological literature was rather obscure. Although widely used in all types of publications, the term was largely absent in the many studies about disasters. Authors who wrote about the effects of concentration camps (Bastiaans, 1970; Volkan, 1979) rarely used the term 'traumatic neurosis' because they assumed it inadequately described the prolonged and far-reaching nature of these effects. It was never utilized in the literature on mourning, in spite of the fact that the loss of a loved one was often described in a manner similar to other traumatic events. The term traumatic neurosis was, however, often mentioned in psychiatric handbooks, but usually only in an oblique manner. The somewhat peculiar position of traumatic neurosis is expressed by the fact that it has never been included in official terminology, such as the three successive versions of the Diagnostic and Statistical Manual (DSM) of the American Psychiatric Association and the International Classification of Diseases (ICD, ninth version) of the World Health Organization.

Transient disturbances

During and after World War II, many other concepts cropped up in the literature. Some focused exclusively on specific events and will therefore be discussed in subsequent chapters of part 2 of this book. In other general concepts the relationship with traumatic neurosis is unclear. The term stress is often employed but in the global sense of unpleasant situations and tension; not in the later, commonly used definitions by Selye, Lazarus and other researchers (see Kleber, 1982).

The category 'gross stress reactions' was included in the first version of the diagnostic manual (DSM-I) of the American Psychiatric Association (1952), to denote all reactions of adults to extreme events. It was part of the group of 'transient situational personality disorders'.

During the '60s, psychological problems resulting from combat were gradually pushed into the background in the Western world. The attention to the effects of natural disasters also gradually decreased. This diminished interest was expressed in the second version of DSM, published in 1968. The group of disorders was now referred to as 'transient situational disturbances' and the term 'gross stress reaction' was replaced by 'adjustment reaction' during adulthood, as distinguished from adjustment reactions during infancy, childhood, adolescence and old age. The change was implemented due to 'reasons of symmetry and classificatory neatness', according to Titchener & Ross (1974, p. 40), who lamented this because they considered 'gross stress reaction' to reflect the unusual circumstances more adequately, while they viewed 'adjustment reaction' as just a pallid term. The new concept referred to the acute reactions of persons without previous disorders to overwhelming strains in the environment. Examples of stressors were: unwanted pregnancy, war, failure at school, retirement (American Psychiatric Association, 1968, p. 49). Contrary to DSM-I, short-lived, psychotic reactions to a shock were also considered as part of 'adjustment reactions'.

The International Classification of Diseases (WHO, 1977), a statistical classification of diseases and disorders, also utilizes categories such as 'acute stress reaction' and 'transient situational disturbances'. Characteristic of these disturbances is the fact that they have bearing upon passing phenomena. How rapidly these phenomena pass is not specified, but the duration is usually no more than a few weeks. Thereafter the problems disappear. If this is not the case, then the diagnosis of another mental disorder is indicated.

Conceptual issues

In reviewing the concepts we have introduced thus far, it appears that several of them indicate traumatic events and their effects in a meaningful, clear manner. At the same time, however, considerable terminological confusion exists in this area. What is regarded by one as traumatic neurosis (e.g. Ladee, 1967), is referred to as an acute stress reaction by another (e.g. ICD-9). Several contradictory views lie behind these various terminologies.

Important questions with regard to the terminology concerning the effects of extreme experiences were, and remain:

- does a concept concern the responses immediately after the traumatic event, or those resulting after a considerable length of time?
- what is the role of existing personality traits?
- do disorders such as traumatic neurosis correspond with other psychiatric disorders (neuroses) or not?
- is it correct to describe the reactions to extreme life events in terms of psychopathology, and thus in terms of mental illness?

With regard to this last question, Bastiaans (1957) points out that a concept such as 'combat exhaustion' (which will be elaborated upon later) was introduced to avoid confusion with the neurosis concept. This term could have bearing upon the traumatic combat exhaustion reactions of healthy soldiers as well as the exhaustion reactions of those who already possessed neurotic personality traits.

Even today these issues occupy an important place in research, as will be shown in subsequent chapters. Clarification of the terminology related to trauma would be possible through a clear delineation of:

a. the immediate, acute reactions;
b. the chronic, long-term reaction patterns;
c. the manifestation of previously latent psychological response patterns;
d. the permanent changes in personality.

Horowitz (1976), however, believes that such a distinction of reactional types does not provide an adequate solution to the diagnostic dilemma. It is difficult to distinguish the four types of reactions in terms of etiology, symptoms or manifestations. Some symptoms occur in all reaction patterns, while personality traits always play a certain role. The solution will therefore have to be sought in more clearly defined concepts.

2.3 Modern developments

Although physicians and technicians already used the term 'stress' in the previous century, it was not introduced as a scientific concept until the '30s and '40s. The Canadian endocrinologist Selye used it to indicate the general physiological adaptation pattern in rats that were confronted with unpleasant stimuli, as he had observed in his laboratory. By the end of World War II, stress research gradually increased, not only in physiology, but also in social medicine and psychology. It generally encompassed physiological research, experimental psychological studies and research into extreme circumstances.

Since the mid-sixties, stress symptoms at work and in other areas of daily life have become an increasingly important focus of research. The influence of all types of events and circumstances upon well-being and health has been and is being studied. Particularly over the past 15 years, stress research has enjoyed an unprecedented boom. The number of publications in this field is vast and continues to increase.

Although many definitions exist, and stress may often be viewed as denoting the entire field of research, the concept of stress is globally described as a disrupted interaction between environmental demands on the one hand and the needs and skills of the individual on the other. This disruption is expressed by divergent physiological and psychological reactions. Cognitive aspects, such as the subjective attachment of meaning to a given situation, play an important role. Furthermore, stress is related to diseases (e.g. cardiovascular diseases) and social problems (e.g. absenteeism); after all, the central focus of stress research is the link between physiological and psychological processes (see Kleber, 1982).

Extreme circumstances have played a minor role in stress research. Nevertheless, along with the boom in stress research, the interest in extreme situations increased. In essence, extreme experiences and their effects form part of the stress field. The issue here is a severely disrupted interaction between individual and environment. Stress and trauma are not two independent concepts but rather two overlapping concepts; stress has a much broader meaning.

As far back as the fifties, Bastiaans (1957) established the link between stress and trauma, when, in his study of concentration camp survivors, he clustered concepts such as homeostasis, adaptation, defense and traumatization around Selye's concept of stress. Stress disrupts the existing balance. As a result, a chain reaction occurs that manifests itself in various syndromes. According to Bastiaans, traumatizing stress is a special form of stress, one with more emphasis on shock, alarm and exhaustion than normal stress situations.

Modern stress researchers concern themselves with the effects of circumstances such as unemployment, conflicts at work, preparation for exams and noise pollution. The 'stressful life events' approach (Holmes & Rahe, 1976), mentioned in the previous chapter, discerns a continuum regarding the degree to which events demand adaptation, and consequently produce an increased risk of health problems. At one end of this continuum are relatively mild events while serious situations are found at the other. The study of traumatic stress focuses on the extreme situations that may produce many or serious stress symptoms. Thus this book will draw upon the theory and research findings concerning stress.

Stress response syndrome

Elaborating upon modern stress research (particularly the work of Lazarus and Janis) and upon psychoanalysis, in the seventies, the American psychiatrist Horowitz introduced the term 'stress response syndrome', which he described as all 'personal reactions when a sudden, serious life event triggers internal responses with characteristic symptomatic patterns' (Horowitz & Kaltreider, 1980, p. 163).

Although Horowitz himself does not provide an exact definition of his concept, we may regard the stress response syndrome as a phasic pattern of coping with extreme events, in which two general categories of response to stressful life events are essential. These two categories are denial and intrusion (see table 2.1).

Denial refers to an intra-psychic process, in which the implications of the event are denied and which is expressed through emotional numbness, the avoidance of thoughts and images of the extreme event, an avoidance of activities reminiscent of what has happened, and a loss of sense of reality. Intrusion refers to the surge of emotions and images that directly or indirectly imply the reexperiencing of the event. Examples of intrusions include nightmares, startle responses triggered by situations similar to the original

Table 2.1 Manifestations of denial and intrusion

Manifestations of denial:
- numbness, not responding to stimuli;
- inability to concentrate;
- amnesia;
- constriction of thinking;
- excessive preoccupation with how things could have happened differently;
- a feeling of not experiencing emotions; feeling numb;
- clinging to old roles;
- continuing to deny that anything has changed;
- loss of sense of reality;
- excessive preoccupation with other matters.

Manifestations of intrusion:
- excessively tense and alert behavior (hypervigilance);
- recurring thoughts about the event and its implications, also during sleep;
- a feeling of being back in the traumatic situation;
- inability to concentrate on other matters;
- sudden surge of emotions concerning the event;
- nightmares;
- looking for the lost person or for elements of the traumatic situation;
- startle responses;
- recurring behaviors linked to the event.

---

event, daydreams, pangs of emotions concerning what has happened and preoccupation with the event. Both of these complementary categories have already been described by Janet and by Freud and Breuer; traumatic events are simultaneously suppressed and repeated involuntarily in the hysterical symptoms.

Basing his conclusions upon research findings concerning disaster, war and grief, Horowitz distinguishes a specific course in the process of coping with a serious event (see Figure 2.1). It often begins with a feeling of bewilderment and disbelief, sometimes accompanied by crying or screaming, and is followed by denial. The individual protects himself from the implications of what has occurred. A striking example is the widow who continues to set the table for her husband, lays out his clothes in the morning and speaks about him in the present tense. In some cases there is no 'outcry' and denial occurs immediately.

Figure 2.1 Phases of the stress response syndrome

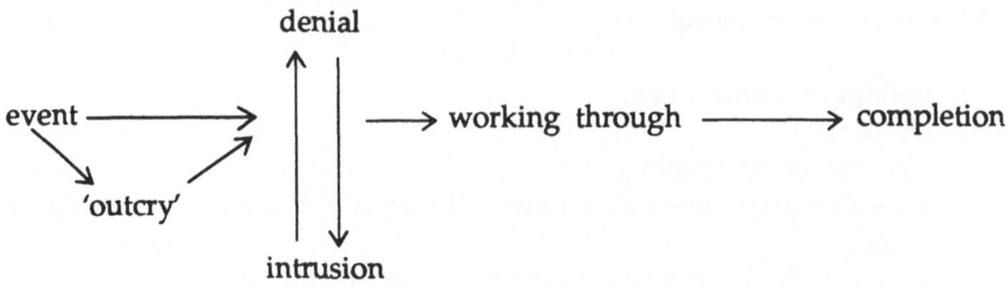

Denial is followed or in many cases accompanied by intrusion. At night the widow dreams about her husband's death. She thinks she sees him on the street and constantly delves back into her memories. Characteristic of the stress response is that denial and intrusion alternate. Horowitz refers to this as 'oscillation'. Sadness and memories alternate with numbness and denial. It is precisely through this alternation that the individual copes with the event. Gradually the event and the related implications are integrated into awareness (Horowitz, 1976b). In the working through or integration phase, less unbidden thoughts and uncontrolled pangs of emotion emerge. The sense of reality increases. Moods become more stable and the significance of the event is more readily accepted. Eventually the completion of this coping process, often distinguished by Horowitz as a final phase, takes place. We will discuss Horowitz' theory in greater detail in the chapters 8 and 12.

The concept of the stress response syndrome does not contend with the previously mentioned problem of neurotic predispositions. No precise distinction is made between symptoms triggered by an event and prior tendencies toward neurotic symptoms, which are manifested as a result of the event. The stress response syndrome encompasses both. Behavior is always an interaction between an individual and a situation, so that symptoms are never produced exclusively by one or the other.

This stress response syndrome emerges after each serious life event. It is a predictable reaction pattern with normal and pathological variations. The pathological stress response syndrome is a coping process that takes longer, becomes blocked (the individual becoming lodged in the avoidance phase, for example), or becomes too intense.

We have discussed Horowitz' ideas because they constitute an excellent example of a detailed theory on coping with extreme stress experiences. Although Horowitz gave the label syndrome to the response pattern following extreme life events and thereby implied symptoms of illness, the theory does not only refer to problematic, psychopathological reactions, but also to the 'normal' psychological effects of extreme events.

A similar view is expressed in the work of Lifton (1979), also a psychoanalyst. He considers the term neurosis unsuitable for what he views as understandable and often justifiable reactions to shocking experiences (such as the case of a soldier who refuses to kill, for example). Lifton, who has become widely known for his study of the survivors of Hiroshima, prefers to speak of traumatic syndrome. Central to this concept is the direct confrontation with one's own death. Although the characteristics of this syndrome are classified and described differently, they are very similar to those of the stress response syndrome. Like Horowitz, Lifton writes about being haunted by images (death imprint), avoidance and numbness and the search for significance in the sudden, shocking changes. Only when these images and other characteristics, such as guilt, prevent normal functioning, does the traumatic syndrome turn into a traumatic neurosis.

Posttraumatic stress disorder

In 1980, the third Diagnostic Statistic Manual (DSM-III) of the American Psychiatric Association was published. Although such a category had been omitted from DSM-II, the DSM-III once again included a separate category to denote the psychological disturbances resulting from extreme life events. There were no reasons given for this renewed entry. However, it is undoubtedly the result of the confrontation with the many problems encountered by Vietnam veterans (chapter 3). Horowitz' work appears to have served as the foundation, although his name is not mentioned.

In 1987 a revised edition of this manual (the DSM-III-R) was published. The five diagnostic criteria for the posttraumatic stress disorder – a category in the group of anxiety disorders – are as follows:

A. The person has experienced an event that is outside the range of usual human experience and that would be markedly distressing to almost anyone.
B. The traumatic event is persistently reexperienced in at least one of the following ways: recurrent and intrusive distressing recollections of the event, recurrent distressing dreams of the event, sudden acting or feeling as if the traumatic event were recurring, intense psychological distress at exposure to events that symbolize or resemble an aspect of the traumatic event, including anniversaries of the event.

C. Persistent avoidance of stimuli associated with the trauma or numbing of general responsiveness (not present before the trauma), as indicated by at least three of the following: efforts to avoid thoughts or feelings associated with the trauma, efforts to avoid activities or situations that arouse recollections of the trauma, inability to recall an important aspect of the trauma, markedly diminished interest in significant activities, feeling of detachment or estrangement from others, restricted range of affect, and a sense of a foreshortened future.

D. Persistent symptoms or increased arousal, as indicated by at least two of the following (not present before the trauma): difficulty falling or staying asleep, irritability or outbursts of anger, difficulty concentrating, hypervigilance, exaggerated startle response, physiologic reactivity upon exposure to events that symbolize or resemble an aspect of the traumatic event.

E. Duration of the disturbance (symptoms in B, C, and D) of at least one month; specify delayed onset if the onset of symptoms was at least six months after the trauma.

The posttraumatic stress disorder is a syndrome that occurs following all types of extreme stressors. The distinction in the 1980 edition between the acute and the chronic type disappeared in the revised edition, DSM-III-R. Note that the posttraumatic stress disorder, by definition, is not an acute stress reaction.

Adjustment disorder

The posttraumatic stress disorder is directly linked to extreme, extraordinary stressors outside the range of normal human experience, such as disasters, wars and accidents. As the next chapters will indicate, however, many of the symptoms listed for posttraumatic stress disorder also occur after other serious stressors, such as the sudden death of a loved one.

For some of these stressors, DSM-III recognizes a different category, 'adjustment disorder,' a barely defined disorder that appears within three months following a psychosocial stressor. Job loss and childbirth are given as examples of such stress factors. Symptoms vary, but usually there is an impairment in social and occupational functioning and/or an excessive reaction to the stressor beyond the normal and expectable level.

While a posttraumatic stress disorder primarily concerns an extreme, unprecedented event, an adjustment disorder refers to a new but less significant situation to which an individual must adjust. We are nevertheless under the impression that both disorders are in line with one another and that several symptoms, such as the intrusions mentioned above, occur in both cases. DSM-III-R, however, does not mention this relationship.

Characteristic for the posttraumatic stress disorder and the adjustment disorder is that both are identified not only on grounds of symptomatology but also on the basis of a causal factor: an extreme situation. This is exceptional, because systems of classification of mental and behavioral disorders normally do not include the etiology of disturbances.

Other stress-induced disorders

According to the extensive classification system of DSM-III-R, there are other pathological responses that may be precipitated by stressful events, although these events are neither necessary nor sufficient to explain the occurrence and form of the disorder. Mentioned are:

- Brief reactive psychosis which is characterized by emotional turmoil and impaired reality testing (indicated by hallucinations, incoherent thinking, delusions or disorganized behavior). The symptoms appear shortly after an unusually serious stressor and last from a few hours to no longer than one month.

- Mood disorders, i.e. major depression and dysthymic disorder (a chronic depressed mood which is not as severe as a major depression, but lasts for at least two years). It is important to realize that symptoms of depressed mood and loss of interest often go hand in hand with a posttraumatic stress disorder.

- Dissociative disorders, such as psychogenic amnesia, depersonalization disorder, and multiple personality disorder. These syndromes are characterized by a partial or complete loss of the normal integration between the functions of consciousness, memory, identity and behavior. There is often a tendency to withdraw from emotionally painful or conflicting experiences, connected with dramatic circumstances such as sexual and physical abuse.

Complications and future trends

In view of the research findings and the theories concerning traumatic events still to be discussed, the posttraumatic stress disorder appears to be a useful category. For example, the two poles of the stress response syndrome – intrusion and denial – are explicitly included in this category. The posttraumatic stress disorder also recognizes, contrary to earlier versions of the Diagnostic Statistic Manual, long-term effects of extreme events. It is clear that this category fulfills a need. Although confusion still surrounds it, within a very short time it has been increasingly used in publications. There is now a wealth of research and clinical papers on posttraumatic stress disorder (Peterson, Prout & Schwarz, 1991).

Objections exist, especially with regard to borderline issues. The first problem is the definition of the extreme stressor. How does one define the first criterion 'outside the range of usual human experience'? For instance, why is the sudden death of a spouse not included in this definition?

Secondly, there is an overlap with other diagnoses of disturbances. Lipkin, Blank, Parson and Smith (1982), for example, included alcohol abuse, psychotic behavior and suicide under the symptoms of posttraumatic stress disorder. Persons whose functioning is impaired after extreme stress may be diagnosed as suffering from more than one disorder. Diagnostic categories are treated as separate entities in DSM-III, but the boundaries between major depression, adjustment disorder and posttraumatic stress disorder are not clear. People report symptoms and these symptoms can be clustered together in constellations which then may be defined as seperate syndromes. However, the empirical basis for such distinct clinical pictures is not firm.

A final disadvantage of psychiatric classification is the lack of discussion of the relationship between the posttraumatic disorder and the 'normal' process of coping with a serious event. A person is considered as diseased or not. The boundary between disordered and normal behavior, however, is not as clear as psychiatric classification suggests. It is a rather arbitrary point on a continuum, and not even a one-dimensional continuum, but a multidimensional continuum.

All these objections refer to the criticism that in a classification system such as DSM-III psychological problems are shackled into a categorical language. It suggests that disorders are discrete entities. This is characteristic for clinical practice, where most decision making is indeed binary (for instance, to treat or not to treat) and where professionals are asked for a straightforward advice by outsiders (for instance, governmental agencies and security companies). We may not forget that the descriptions and criteria of DSM-III were developed by consensus in committees of clinicians (mainly psychiatrists) rather than by reference to an existing body of empirical language. The various nuances and subtleties of empirical studies using continuous measures and component dimensions are not satisfactorily represented in the diagnostic categories of psychiatric classification. 'Psychological problems are real, but are not entities. They are not discrete. They are not something that is entirely present or entirely absent, without shades in between' (Mirowsky & Ross, 1989, p. 11).

It is now debated whether reactions to severe stress are to be considered a subdivision of anxiety disorders, as in the DSM-III-R, or a new subdivision of dissociative disorders, or even a distinct group of disorders. In the proposed tenth version of the International Classification of Diseases (ICD-10), a special class of disorders related to stress is included. It consists of 'posttraumatic stress disorder', 'adjustment disorders' and 'acute stress reaction', a disorder with an immediate onset after an exceptional stressor. ICD-10 also proposes the diagnosis of 'enduring personality changes after catastrophic experiences', a concept that refers to specific and permanent changes in personality after long-lasting extreme situations (chapter 6).

2.4 Conclusions

Over the years, several concepts have been developed to indicate the effects of extreme experiences. A researcher's or clinician's preference for certain concepts is contingent upon the 'Zeitgeist' of his time, his academic discipline, theoretical background and personal views. Which concepts do we wish to use at this time? Here is a summary of the most significant concepts discussed in this chapter.

The term with the broadest meaning is stress, but it is too general to be used exclusively for an analysis of traumatic events. Furthermore, it refers more to the experience itself than to, for example, long-range effects. The term is suitable as a general denotation of the entire field of study – the study of extreme stress situations and their effects. Furthermore, it clearly indicates that there are links to other fields in which stress symptoms (such as those occurring in the work place) are being studied.

We want to be cautious with the term 'trauma'. A considerable number of those involved in researching extreme situations do not use this term. Trauma is sometimes understood as physical injury, a definition that is common within the medical arena. The alternative concept of psychotrauma has scarcely been established. More important, however, is the objection that this concept suggests a gap between people who have problems and people who do not: one person has a trauma, another does not. Utilizing the concept in this manner readily leads to reification, i.e. trauma is viewed as a separate entity that can be seen independently from its context, as in the case of a heart attack. Although this danger exists, in principle, for every concept, it especially applies to the term trauma. After all, the term is used to indicate an event as well as the results thereof. All of these objections do not apply to the content of the original concept of trauma as used in psychoanalysis: the Freudian concept is certainly useful and indicates important aspects that will be discussed in subsequent chapters. Our caution with the term trauma results from the confusing usage of the term in publications other than the original ones.

Similar considerations apply to traumatic neurosis, which has already been discussed extensively above. It remains curious that the usage of this term has been so marginal and is being used less and less today.

We prefer concepts such as coping or stress response syndrome, in which normal and pathological responses are not separated, so that the normal adaptational process is included in a theoretical and research approach. Every individual suffers from symptoms after experiencing an extreme life event. Only the severity and duration – and the necessity of professional help – can vary substantially, depending upon the influence of various social and personality characteristics.

Coping with a severely distressing event may for some reason not be successful: coping is blocked or the symptoms are too severe. One of these maladaptive responses to severe stress is the posttraumatic stress disorder, the new manifestation of the concept of traumatic neurosis. Other disorders are brief reactive psychosis, dissociative disorders, mood disorders, and adjustment disorders. We will not attempt to differentiate between these different disorders, as the dividing lines between them are usually not very evident. In general we will refer in this book to coping processes and to disorders in coping strategies.

PART II

Traumatic Events

CHAPTER 3

WAR

This chapter deals with the psychological problems of soldiers resulting from acts of war. Before we review the research findings about combat stress, the history of terminology will be described. The following section examines the research findings from World War II and subsequently those from the Vietnam war. In the latter case, the postwar effects in particular have demanded attention. Research on this subject has greatly influenced the modern field of traumatic stress studies.

Psychological disorders resulting from combat were probably first identified in modern science in the American Civil War with reports of the so called 'soldier's heart' – coronary symptoms due to the strains of battle – by DaCosta (Kellett, 1982).

During World War I, the term 'shell shock' was introduced by Mott (Trimble, 1981) to describe the psychiatric symptoms resulting from violent acts of war. The term assumed the effects to have an organic basis: the patients' brains had been hit by fragments of grenades and bombs. When, however, it was established that emotional disturbances could occur even in soldiers remote from any exploding missile, the term 'shell shock' was no longer used (Brill, 1967). It was felt that psychical causes were involved (Trimble, 1981).

In 1919, Freud employed the term 'Kriegsneurosen'. He considered war neuroses to be traumatic neuroses, resulting from a conflict in the ego: a conflict between the soldier's old, peace-loving ego and his military ego. This conflict would be absent in traumatic neuroses in times of peace. Freud's definition of the term 'war neuroses' is not very clear.

Freud's work on war neuroses was not exceptional. Many publications, especially in French (e.g. Roussy & Lhermitte, 1917) and German (e.g. Hübner, 1917), were dedicated to the psychological disturbances of soldiers during and after the fighting in the trenches.

Knowledge acquired during World War I was largely forgotten by the start of World War II. This is indicated by the fact that approximately 90% of the patients that military psychiatrists had to deal with, could not be classified with any of the existing categories (Finkel, 1976). Moreover, the symptoms did not fit into the traditional view of neurosis (Brill, 1967).

A special committee of American psychiatrists which toured Europe to analyze psychiatric problems during World War II, concluded that the symptoms could not be described in terms of neurosis, nor as simply a state of exhaustion, nor as anxiety. Their final conclusion was that it was 'a temporary psychological disorganization out of which various more definite and more familiar syndromes evolve' (Bartmeier, 1946; quoted by Kormos 1978, p. 5). It was due to these diagnostic difficulties that a new diagnostic system was developed (chapter 2).

In the literature about World War II the term 'war neurosis' is used as a global concept that refers to all abnormal reactions of soldiers in wartime (Brill, 1967). A distinction is made between 'gross stress reactions' (i.e. symptoms resulting from experiencing violent acts of war) and prior neuroses that have surfaced as a result of the violence. The term (acute) traumatic neurosis is usually equated with 'gross stress reactions'. This division appears in DSM-I.

Another distinction was made between war neuroses caused by the battlefield with all its misery, and the problems of soldiers caused by factors not associated with the immediate violence of war (being far away from home, being a soldier, etc.). Little has been written about the latter category. The impression is clearly created that the former problems – summarized by the term combat exhaustion – occurred far more frequently.

The phenomenon of combat exhaustion – the breakdown of a soldier after his having fought at the front for a considerable length of time – will be dealt with extensively in this chapter, as research on this subject was the starting point for the study of other extreme experiences. Many concepts, assumptions and problems that were first formulated here, are still relevant to modern research.

One manual dating from that period (quoted by Archibald & Tuddenham, 1965) listed the following five criteria for combat exhaustion:

1. overwhelming stress must precede it;
2. the individual had a 'normal' personality prior to the condition;

3. the disorders are curable;
4. it is possible for the symptoms to develop in the direction of a traditional neurosis;
5. if the problems persist, combat exhaustion should be used as a temporary diagnosis until a more definite diagnosis can be made.

This last criterion indicates that psychiatrists and psychologists did not know how to handle prolonged stress reactions to violent acts of war. Long-term effects did not fit well into the concepts of that time and were assumed to be related to already existing neurotic disorders.

The concept of combat exhaustion became again relevant in the seventies and eighties. Kormos (1978) defined acute 'combat reaction' as the behavior of a soldier under conditions of war, which is interpreted by those around him as a sign that he has stopped functioning as a soldier. The communicative meaning which is given here to problems of war is noteworthy. A soldier transmits a 'message' to his environment that he can no longer function as a soldier. At the same time he no longer *wants* to function. Psychological reactions to combat have more recently been examined by Solomon and her colleagues (1987) in a study of Israeli soldiers who fought in the 1982 Israel-Lebanon war. She too defined combat stress reaction as a condition in which a soldier ceases to function militarily and acts in a manner that endangers himself and his fellow combatants.

3.1 The effects of combat exhaustion in World War II

In a classic article, Swank (1949) described the phenomenon of 'combat exhaustion' and its related symptoms. He studied data on more than 4500 soldiers (mostly Americans) who had broken down after prolonged battle in Western Europe. These were young, motivated, healthy soldiers who had successfully passed several selection procedures.

The syndrome of combat exhaustion may be outlined as follows. After a, sometimes very long, period of adequate adaptation to the circumstances of war, a soldier begins to become fatigued. This fatigue is chronic and does not disappear easily. The soldier becomes unable to distinguish between the many sounds of war. He is easily confused, becomes jumpy and loses his self-confidence. Irritation occurs with increasing frequency. The soldier overreacts to all kinds of stimuli; for example, he immediately drops to the ground at the slightest sound. His features betray anxiety. Finally the soldier becomes apathetic and passive. He cannot remember details. He becomes sluggish both physically and mentally. This condition could be dangerous;

some soldiers walk straight into enemy lines. The entire syndrome is accompanied by physical reactions, feelings of inadequacy, fear of crowds and a desire to be alone.

Swank (1949) distinguishes four groups of symptoms that emerged in the soldiers he studied, all of whom had been removed from the battlefield:

I Emotional tension and related symptoms: fatigue, anxiety, insecurity, fear of groups, tendency towards isolation, fear of failure, fear of losing comrades, irritability, emotional exaggeration (laughing loudly, crying), sleeping problems (very common), nightmares, startle responses, increased alcohol consumption, reduced appetite.

II Cognitive problems: loss of memory, apathy, confusion, preoccupation with certain things (the fighting experiences, for example), loss of spontaneity, loss of concentration.

III Physical problems: headaches (the most frequent somatic reaction that occurred in many varieties), dizziness, stomachache, backache, trembling, rapidly emerging fatigue and many other symptoms.

IV Hysterical manifestations: stammering, amnesia, tremor and paralyses of the limbs. These symptoms rarely occurred (in 4% of the cases).

Swank's research clearly shows that the nature and degree of the symptoms were remarkably uniform for all soldiers. Although this could have been caused by the similarity of the selected soldiers or identical circumstances, the author believed that the symptoms were a normal, specific reaction to a prolonged, severe stress factor. The similarity with Selye's findings concerning the phenomenon of stress – also a specific pattern, caused by non-specific stimuli – is striking.

Other authors are less convinced of this uniform pattern of symptoms. They all mention more or less the same reactions, but argue that one soldier may experience more fears and memory disorders, while another leans toward somatic reactions. Therefore, the occurrence of symptoms varies for different soldiers (Kormos, 1978).

In their classic book 'Men under stress', Grinker and Spiegel (1945) point out that soldiers exhibited every possible psychiatric and psychosomatic symptom. Combat exhaustion is characterized by a multitude of symptoms. It is not possible to reduce this multitude down to different types of exhaustion.

Despite this multitude of symptoms, also Grinker and Spiegel conclude that the symptomatology for soldiers is rather uniform: 'the identical symptoms occur in various permutations in almost all cases' (1945, p. 219). Although the intensity and combinations of reactions may differ, the symptoms that emerge are mostly the same. We therefore conclude this section with the list of the 10 most frequently occurring symptoms that Grinker and Spiegel observed in their patients in the military hospital, and that emerge as being important in recent studies (Kalman, 1977):

1. restlessness;
2. irritability and aggression;
3. fatigue on arising and lethargy;
4. difficulty falling asleep;
5. anxiety;
6. frequent fatigue;
7. startle reactions;
8. feeling of tension (e.g. vomiting);
9. depression;
10. changes in personality and memory loss.

Acts of war, neurotic personality or interaction?

A vehement debate took place in the forties concerning the causes of psychological problems of soldiers, as discussed earlier in chapter 2. Many psychiatrists and psychologists initially assumed that already existing personality traits were at the root of combat exhaustion. They based this assumption on psychoanalysis. On the other hand, those who actually conducted research among soldiers concluded that the symptoms were more closely related to the nature and gravity of the acts of war. As usual, the truth turned out to lie somewhere in between. Both situational and personal aspects determined the emergence of problems, although already existing personality traits appeared to occupy a rather marginal position.

Swank (1949) established in his extensive study the degree to which collapsed soldiers possessed 'pre-combat neurotic and psychopathic traits'. In general, it was found that soldiers with few neurotic traits could hold out somewhat longer on the battlefield than those who clearly exhibited neurotic traits (an average of 36 days as compared to 30). The differences between both groups primarily occurred in breaking down very quickly or becoming exhausted only after a very long period of fighting. A large part of the so-called unstable soldiers remained active in the army for considerable periods of time.

Furthermore, it was established that the nature and degree of symptoms were barely related to pre-existing neurotic personality traits. A few physical symptoms and increased emotionalism occurred somewhat more frequently in unstable soldiers.

Combat exhaustion, on the other hand, turned out to be dependent, above all else, upon the number of victims in the fighting unit. The larger the number of soldiers who were wounded, killed or taken prisoner, the more frequently combat exhaustion occurred. Based on his own research as well as other studies, Swank concluded that most soldiers begin to exhibit symptoms of exhaustion when approximately 65% of the original military unit has fallen victim to the acts of war. This applied both to units where this percentage had already been reached by the tenth day of fighting, and to units where it took 90 days. In the former groups the combat exhaustion was accompanied by much more emotionalism; in the latter, on the other hand, the element of apathy occurred with more frequency. The nature and degree of symptoms were thus related to the nature and intensity of the fighting. In general, it was established that the soldiers who had fought the longest before breaking down, showed a greater number of symptoms and more severe symptoms. Other researchers reached the same conclusions (Grinker & Spiegel, 1945).

Many researchers were surprised by the fact that soldiers who were considered unstable held out so long during the war. Rachman (1978) mentions a study by Hastings, Wright and Glueck of 150 successful air force soldiers who all had experienced heavy combat. Approximately 50% of these soldiers came from emotionally unstable families, and were unstable themselves.

Caution is required when drawing conclusions about personality predispositions. Assessment of such traits was not optimal in those days, because there were no diagnostical instruments, or because measurement was only retrospective. In addition, no random sample of the population was used; the soldiers were selected army men. Furthermore, it only involved American soldiers, i.e. well-fed, well-trained and motivated men from a country where no war was taking place. Unfortunately, very little research material is available on German, British or Russian soldiers.

The conclusion that pre-existing personality traits are irrelevant to the occurrence or non-occurrence of combat exhaustion is rather exaggerated in the light of a number of studies that contradict this statement. Brill and Beebe (1955) discovered that the chances of a breakdown were seven to eight times higher for men with prior neuroses than for mentally well-adjusted soldiers. These authors considered the 'pre-service personality' to be the most important factor in combat exhaustion. But they too cautioned against the assumption that everybody with emotional problems would break down during battle.

Predisposing personality traits are related to the occurrence of combat exhaustion, in the same way as variables such as age, education and marital status (see Brill, 1967). They, however, do not constitute the most important causes of it. Rather, all these variables mediate the effect of combat. Combat exhaustion is best considered to be the result of an interaction between individual and situational aspects, in which all aspects need not be equally important. This thought is very much present in the work of Grinker and Spiegel (1945), who oppose an exclusive emphasis either on individual predispositions or on war circumstances. Soldiers with problematic backgrounds will break down sooner under pressure, but at the same time it should be clear that eventually every soldier breaks down when confronted with an extremely overwhelming situation. It must not be forgotten that this interaction is of a dynamic nature.

Frequent or infrequent occurrence in World War II?

How frequent was mental breakdown? This question may appear somewhat futile, when one realizes that the chances of a soldier surviving were often no higher than 50% and on some battlefields considerably lower.

General figures are available. In the American Civil War, when the phenomenon of combat exhaustion – sometimes indicated as nostalgia – was first observed, and in other wars during the 19th century, an average of two to three soldiers out of every 1000 collapsed each year (Kellett, 1982). During World War II, however, the number was considerably higher. Between 28 and 250 out of every 1000 soldiers were admitted to hospitals for psychiatric reasons (Bourne, 1970; DeFazio, 1975). Kellett (1982) mentions that psychiatric cases amounted to between 7% and 20% of the total number of victims. Combat exhaustion concerned a minority which, considering the size of the battle, reached considerable proportions. A total of 850.000 Americans were admitted to army hospitals during World War II for so-called neuropsychiatric reasons (Brill, 1967).

It should be noted that the majority of the soldiers suffered from complaints and symptoms resulting from the violence of war. Only 1% of 4500 soldiers who were questioned, denied ever experiencing anxiety during the fighting; 33% to 50% reported they experienced anxiety during nearly every mission (Schaffer, 1947; quoted by Rachman, 1978). All kinds of physical expressions of fear, as well as nightmares, occurred frequently.

Why did soldiers break down at a certain point? As reported earlier, Swank (1949) discovered that soldiers became mentally exhausted when approximately two thirds of their comrades had been killed, wounded or taken prisoner. They were confronted with death to such an extent that they broke down. The fear of being killed or wounded themselves played an important part, along with experiences of extreme desolation. However, combat exhaustion could not be attributed exclusively to this. After all, soldiers who had serious breakdowns were apathetic and seldom ran away. Stories about soldiers who walked into enemy lines in a daze are well-known. It was the total uncertainty about the moment of death and the resulting pressure, which led to the breakdown (Grinker & Spiegel, 1945). The aspect of loss of control is also found in modern literature (chapter 8). This is in accordance with the conclusion that the breakdowns occurred especially in situations of prolonged attack by enemy troops, while they themselves were unable to strike back (Bourne, 1970).

Anxiety and uncertainty were best controlled by trust, according to the earlier mentioned study conducted by Shaffer (quoted by Rachman, 1978). Trust in equipment, in the rest of the men and in the officers were crucial in coping with the anxiety. Promotions, salary increases, and ideological convictions were not very effective.

This section and previous ones show the relevance of control. It is important that the individual feels in control of the situation, that he feels he can cope with it, that he possesses sufficient skills and that he trusts his equipment and his fellow soldiers. All of these aspects are closely related to the group the soldier is a part of. In scientific literature on war, the importance of the small group for motivated perseverance in battle is emphasized innumerable times.

Treatment of combat exhaustion

During World War II, psychological combat exhaustion was treated according to the principles of 'immediacy, proximity and expectancy'. As soon as possible after the problems emerged, the patient had to be treated as close as possible to the front, in the expectation that he could return to combat quickly.

This treatment was based on scientific as well as on practical and ideological considerations. This latter aspect is expressed in the training of the attending staff. They were trained to believe that mental breakdown was a 'normal and reversible' reaction to war (Kormos, 1978). After all, every soldier was needed in combat. Glass (1954) and Van Meurs (1955) stated that a soldier, evacuated because of psychological disturbances, would probably receive so much support from others for his inability or unwillingness to fight, that it would later become impossible to induce him to fight.

This line of reasoning did have an empirical basis. Both World Wars showed that symptoms quickly begin to lead lives of their own. In his research on the British-Japanese front, Williams (1951) discovered that symptoms became more permanent and the prognosis of returning to combat became worse as the patient was taken further away from the front. In recent years Solomon and Benbenishty (1986) proved that intervention along the lines of 'immediacy, proximity and expectancy' was beneficial in treating soldiers with combat stress reactions resulting from the Israel-Lebanon war.

Although this view of immediacy, proximity and expectancy of fast recovery was developed during World War I, it is not clear whether it was applied frequently at that time. The assumption was that a soldier who had a mental breakdown, was lost to the battlefield for good (Bourne, 1970). The methods of treating psychological combat exhaustion used during World War II are listed in table 3.1.

Table 3.1 Treatment methods for combat exhaustion in World War II

- Rest, often with the aid of sedatives and relaxation.
- Narcotherapy. A mild narcotic was administered to the soldier in order to increase his sensitivity to suggestions. He was then told to talk about what was bothering him and to express his feelings. Narcotherapy was used very frequently during World War II because it yielded results relatively fast.
- Hypnosis, particularly in combination with narcotherapy.
- Verbal therapy, a form of usually very directive talks about the soldier's traumatic experiences.
- Group discussions and group therapy, in which the group consisted exclusively of patients.
- Sleep therapy lasting several days.

Treatment of psychological problems during and immediately after World War II should not be viewed simply in terms of our present criteria. After all, there was a war going on. Long-lasting, non-directive or extensive therapies are somewhat out of place in a field hospital amid the turmoil of battle.

In general, the results of treatment were perceived very positively during the forties. Swank (1949) concluded that an adequate medical treatment, one which did not have to last long, was capable of eliminating all symptoms, even if they had existed for a considerable time.

However, there was no thorough, reliable evaluation of these successful results. Kormos (1978) questioned the percentages of recovery that were given in the forties. Soldiers who returned to battle and who were subsequently wounded or killed were considered as soldiers who had successfully recovered from combat exhaustion.

3.2 The years following World War II

Up to this point we have discussed only those stress reactions that emerged during or shortly after the acts of war. Is it possible, however, that prolonged confrontation with these extreme, life-threatening circumstances results in long-term effects? Long-term refers to the period of leaving the service and returning home, the period immediately following the war and the postwar years.

Nowhere does the literature from the forties even imply that soldiers may experience problems in the long run. It is therefore not surprising that little research has been conducted on the long-term effects of psychological combat exhaustion and the fighting during World War II. No more than a handful of publications on this subject have appeared.

Whether soldiers experienced problems immediately after World War II is not really known. This question would certainly not have been very popular at that time. After all, the soldiers were celebrated as heroes, and heroes are not supposed to have mental problems. It might be that indeed these soldiers had few posttraumatic effects, as is sometimes suggested in modern literature on Vietnam veterans (DeFazio, 1975, 1978).

Brill and Beebe (1955) conducted a follow-up study on 1000 soldiers who had been hospitalized due to psychiatric problems after the war. Five to six years later, 90% of them suffered from symptoms, particularly irritation, anxiety, gastrointestinal disturbances, restlessness and headaches. Most veterans reported that these symptoms had first emerged at the time of their breakdown. They believed their health to be slightly worse than at the start of their service, although better than at the end of it. Approximately 40% believed they still needed treatment; 14% were unable to hold down a full-time job, due to the problems. More than 60% of the veterans had no diseases or neurotic problems. Although Brill and Beebe found an improvement since the end of the war, their findings clearly suggest that the problems had by no means disappeared completely after the war, even with early treatment.

Archibald and his colleagues (Archibald, Long, Miller & Tuddenham,

1962) compared war veterans who had suffered breakdowns during the war, with veterans who did not participate in combat during World War II, 15 years after the war had ended. Both groups had been selected at random from the files of a Californian hospital.

The war veterans reported various anxiety symptoms to a significantly higher degree than the 'non-combat controls': sleep problems, dreams about fighting, dizziness, perspiring, being jumpy, giving up hobbies and favorite activities, avoiding war movies and noise. The veterans were in a chronic state of hyperalertness, and they considered their current problems to be more serious than during or immediately after World War II. However, the control group reported much the same. Thus, war problems do not necessarily become worse over time; people may more readily complain about the present than about the past.

Also remarkable was that the veterans who had collapsed in the past, had more frequently come from the lower social classes than the veterans in the control group. Furthermore, more of them had lost one parent before they reached the age of 14, particularly in early childhood. Other data showed an individual history of deprivation (this does not mean that this history is responsible for later problems; in every war, people from the lower social classes and with less education are more likely to be placed in the combat zones).

As in most studies on extreme experiences, Archibald et al. (1962) found hardly any psychotic symptoms in war veterans. Moreover, there were very few personality differences between the two groups.

In a follow-up study, Archibald and Tuddenham (1965) concluded that 70% of those who had suffered a so-called acute traumatic neurosis during World War II still had a chronic traumatic neurosis after 20 years. One third of these war veterans was unemployed and one third was in an uncertain employment situation. Nevertheless, the question remains whether many of the soldiers on World War II experienced war-related problems later on. After all, the subjects in the studies mentioned above were admitted to the hospital because of psychological complaints.

We cannot conclude how many soldiers suffer from prolonged symptoms and problems resulting from battle activities and how severe these symptoms are. It is certain that a number of soldiers still experience disturbances that disrupt their daily lives many years later.

3.3 The Vietnam war and its effects

The psychological and social effects of the Vietnam War on American soldiers provided the impetus for the development of the diagnosis of 'post-traumatic stress disorder' in the new psychiatric classification (see previous chapter). As a matter of fact, the psychological casualties of the Vietnam war were for a large part responsible for the renewed interest in traumatic stress.

The Vietnam War was the longest war in the history of the United States. Approximately 3.2 million men and women served in the American army in Vietnam and more than 50,000 were killed. The fighting in Vietnam clearly differed from the engagement of the United States during the two world wars. The United States were internally divided about the usefulness of American interference in Vietnam. Moreover, the motivation of the soldiers themselves was often rather weak. Aside from that, combat in Vietnam consisted of guerilla-warfare in the jungle, a type of warfare western soldiers are not trained for. Finally, high drug abuse by American soldiers complicated military functioning.

Over the last years many scientific studies about Vietnam veterans have appeared in psychology and psychiatry. In 1982 the number of publications was already more than 160 (Fairbank, Langley, Jarvie & Keane, 1981; Silver, 1982). Till 1991 more than 500 articles had been published in psychological journals.

During the Vietnam war

At first, many psychological problems were expected to emerge during the war, especially because of the special circumstances and the opposition in the United States. Halfway through the war it did not turn out to be as bad as expected; between 12 and 15 men out of every 1000 were hospitalized for neuropsychiatric reasons (Bourne, 1970; DeFazio, 1975). Six percent of all medical evacuations from Vietnam concerned psychiatric cases; in World War II this number was 23%.

Relatively few cases of psychological combat exhaustion occurred. The diagnosis of 'combat fatigue' was made in 15% of all psychiatric hospitalizations (Bourne, 1970). Personality and behavioral disorders, on the other hand, amounted to 40%. Various military and psychological causes were believed to explain the relatively low occurrence of psychological problems: the soldiers were in action for only one year, and they had sound psychological preparation, much rest, short-lasting missions and a lot of relaxation.

Bourne suggested that the psychological problems that occurred in Vietnam were related to pre-existing psychopathology.

After a while, the successful arrangement of an one year active duty period, also appeared to have negative effects. They prevented, to a great extent, the development of a primary group to which the soldier felt he belonged and which he identified with. The dates of arrival, departure and relaxation periods were known in advance. On the plane the soldier sat next to unknown soldiers; during combat, old comrades left and new soldiers arrived. The solidarity of the small unit that proved to be so important during World War II disappeared and was replaced by an individual morality. The soldier was forced to solve his own problems, and deal with his own doubts about the war and his own involvement in it.

The first studies

The literature on World War II focused on the psychological problems occurring during and shortly after the combat activities. The long-term effects of the war were barely discussed. Oddly enough, the opposite holds true for the literature on Vietnam. Little research was undertaken during the war. Only after it ended (1975) publications appeared, a little hesitantly at first (Horowitz & Solomon, 1975), but in increasing numbers later on. These studies were primarily concerned with the effects of the war several years after the soldiers returned to the United States. This is quite striking. Midway through the seventies, it was believed the misery of Vietnam was finally over.

Nace and his colleagues (Nace, Meyers, O'Brian, Ream & Mintz, 1977) studied some 200 veterans (including many drug addicts), who had returned from Vietnam an average of 28 months earlier. Approximately one third of them belonged to the depressed category, as measured with the Beck Depression Inventory. 'Normally' this percentage is somewhere between 15 and 23%. The non-depressed group hardly differed from the depressed group on variables concerning service in Vietnam. Unemployment was considerably higher for the depressed group, as were the number of divorces and violations of the law. Both during and after the Vietnam war, the depressed group's consumption of heroin and barbiturates was much higher. Conclusions about causal relations with respect to these findings are difficult to draw. Depression may be caused by the Vietnam war, but also by unemployment and divorce as well.

Helzer et al. (Helzer, Robins, Wish & Hesselbroch, 1979) also studied depression (measured through interviews) in Vietnam veterans. An initial study showed more severe symptoms of depression 8 – 12 months after the war in a veteran group in comparison with non-veterans. Two years later, the differences between these groups disappeared.

These findings by Helzer et al. are an exception. Most researchers found significant effects even several years later. Figley (1978) reports a study which showed that soldiers from combat units had more nightmares, fewer friendships and a lower morale than soldiers who had not fought in Vietnam (see also Helzer et al., 1979). But when variables such as education, deviant behavior during adolescence and parental care were controlled for, these differences disappeared to some extent.

Compared to veterans who had not been in Vietnam, the Vietnam veterans experienced twice as many problems in the form of drug and alcohol abuse, psychological tensions, nightmares and injuries (Ewalt, 1981). Vietnam war combatants experienced more problems than non-combat Vietnam veterans. The problems of the combat soldiers did not disappear; for other Vietnam veterans the severity of their symptoms gradually decreased (Ewalt, 1981). As was shown in the studies by Archibald and colleagues, there were no differences between these groups with respect to psychoses.

It is striking that the consequences of war also extended to the social arena. Problems emerged within the family and at work, and many veterans complained about feelings of alienation from society and its institutions (Lipkin et al., 1982).

The current picture

Let us return to the question of the extent to which veterans or soldiers will suffer from problems. Research on World War II barely provided an answer, as only those people who had turned to a hospital as a patient were studied. The studies on Vietnam veterans provide a better answer to this question, because they sometimes employ random samples of Vietnam veterans. Two good examples of well designed studies are the one by Frye and Stockton (1982) and the one by Kulka, Schlenger, Fairbank, Hough, Jordan, Marmar and Weiss (1990).

Frye, a Vietnam veteran himself, asked his former military classmates who had fought in Vietnam to answer several questionnaires. This enabled him to obtain a group of 88 Vietnam veterans from the period between 1968 and 1971.

In terms of various background variables, this group emerged considerably more favorably than the soldiers from the other studies mentioned in this section. Nearly all attended college, hardly anyone was unemployed at the time of questioning and 83% of the veterans earned an excellent income. As they were slightly older than the average Vietnam draftee, one might expect they would experience fewer problems so many years after the war.

Frye and Stockton used a questionnaire to discover whether these veterans suffered from a posttraumatic stress disorder, as defined in the DSM-III classification. The questionnaire contained questions about the three diagnostic criteria of reexperiencing the traumatic event, the avoidance of involvement in all sorts of Vietnam-related matters and emotions, and the presence of certain symptoms that had not been significant in the individual's life prior to Vietnam.

It was established that 23.9% of the veterans showed the syndrome and that 19.3% were on the borderline (two criteria applied very strongly for an individual and one to a somewhat lesser degree). Ten years after discharge from the service, a considerable minority showed clear sytmptoms of a blocked process of coping with the traumatic experiences. Veterans still suffered from nightmares, intrusive thoughts, listlessness, feelings of guilt and an awareness of loneliness.

It is worthy of note that few cases of combat exhaustion or traumatic neurosis were diagnosed during the Vietnam war, but that the symptoms of posttraumatic stress disorder only emerged after the soldiers had been back in their country for a considerable length of time!

The comprehensive National Vietnam Veterans Readjustment Study was a very ambitious and sophisticated study of the postwar psychological problems of veterans (Kulka et al., 1990). Its primary goal was to provide information about the prevalence, incidence and effects of posttraumatic stress disorder (PTSD) and related postwar problems in readjusting to civilian life. The sample of the study consisted of 1600 Vietnam theater veterans and was nationally representative. Adequate comparisons groups were included to provide a context for understanding the adjustment problems of veterans.

It was found that 15.2 percent of all male Vietnam veterans fullfilled the criteria of PTSD. This rate was much higher than the rates for comparable veterans who had served in the U.S.A. (1.2 percent) or civilian counterparts (1.2 percent). The current prevalence of PTSD among women who served in Vietnam was estimated at 8.5 percent. Another 11.1 percent of male and 7.8 percent of female veterans suffered from clinically significant stress reaction symptoms that adversely affected their lives but were not of the intensity or breadth required for a diagnosis of PTSD. These findings indicate that approximately 836,000 veterans in the United States are current cases of complete or partial PTSD.

The findings of this national study indicate a strong relationship between posttraumatic stress disorder and other postwar readjustment difficulties. Male veterans with PTSD are two to six times more likely to abuse alcohol or drugs than those without the disorder. The lifetime prevalence of alcohol abuse of the veterans was nearly 49 percent. Veterans with PTSD are five times more likely than those without to be unemployed; almost one fourth are currently separated; 70 percent have been divorced; and 37 percent had committed six or more acts of violence during the past year. Having PTSD was associated with many other problems in virtually every domain of these veterans' lives.

Past and present readjustment was found to be most difficult for veterans who were most heavily and directly involved in the war. Veterans with high levels of exposure to combat and other war-zone stressors had considerably more psychological problems than veterans with a low or moderate exposure. There were also substantial differences between ethnic categories. The differences between white and black veterans could be attributed to the fact that blacks were more likely to be exposed to war-zone stress, but the even higher prevalence of PTSD among Hispanic veterans (twice as high as the rate among whites) could not be explained by this factor.

This national study is truly impressive, although it does not deal at all with the *process* of coping with the aftermath of the Vietnam war. It provides overwhelming evidence for the existence of posttraumatic stress disorder and other long-term consequences among veterans.

More disturbances than in other wars?

With some caution, we may conclude that veterans from the Vietnam War showed more problems than veterans from World War II. This difference may be related to the nature of the war in question. The Vietnam war was a fight in the jungle, an unknown terrain for most Americans. The enemy was rarely or never seen. The enemy soldiers were part of the native population; it was unclear to the American soldier which side this population was on. And finally, the culture and language of the Vietnamese were very different from their own. However, all these factors do not sufficiently explain the emergence of problems after the war, if only because during combat in Vietnam, i.e. during the confrontation with the aspects of war mentioned above, the psychological disturbances did not frequently occur.

An additional and perhaps more satisfactory explanation for the many problems of Vietnam veterans is the social climate they encountered upon their return to the United States. During the course of the war, public opinion increasingly turned against military intervention in Vietnam, so that the discharged soldiers were not received with enthusiasm. This trauma of returning home will be discussed further in the chapter on determinants (chapter 9) as it clarifies the role of social factors in the effects of traumatic events.

Although several research findings point in the direction of a higher number of psychological disorders in the case of Vietnam veterans than in veterans of other (American) combat activities, this has yet to be proven. Thienes-Hontos, Watson and Kucala (1982) studied the occurrence of stress symptoms of posttraumatic stress disorder in Korea and Vietnam veterans. There was little difference in the frequency of the stress symptoms, established according to DSM-III criteria. Ten percent of the Korea veterans suffered a full-fledged posttraumatic stress disorder, and 7% of the Vietnam veterans did so; on the borderline were 27% and 17%, respectively.

These findings undermine the assumption that Vietnam veterans have more problems. On the other hand, they also evoke interesting questions. These figures, which refer to an average period of two years after returning from the war, are surprisingly lower than those found by Frye and Stockton for veterans 10 years after the war. In the latter group, 43% suffered from the syndrome (complete and borderline), whereas the figure is only 24% in the study mentioned above; Thienes-Hontos et al. studied hospital patients and Frye and Stockton studied a group of reasonably to very successful veterans. Both studies employed the same criteria. This suggests that the effects of the Vietnam war have become stronger in the long run. Such a conclusion underlines the importance of the determinants of the coping process (including the social situation), which play an important part after the soldiers have returned home.

3.4 Conclusions

In this chapter we discussed two closely related phenomena: psychological breakdown during war (combat exhaustion) and the prolonged effects of acts of war (posttraumatic stress disorder). Research on these phenomena has greatly influenced the entire field of traumatic events.

One could argue that the violence of war, because of its long duration, is not the most appropriate example of a traumatic experience. Yet it is pre-eminently characterized by powerlessness, disruption and discomfort. War may be regarded as a succession of extreme situations that overwhelm the individual and fundamentally affect his feelings of security. The soldier is confronted with lack of sleep, noise, physical hardship, fear of death, separation from his familiar environment, lack of normal living conditions and the death of his companions. The powerlessness which results from these circumstances may lead to combat exhaustion. The disruption will haunt the soldier later, in times of peace. The misery and cruelty of war may seriously damage his confidence in himself and in the world and may isolate him from his environment.

World War II literature focused especially on combat exhaustion, the psychological and physical breakdown of a soldier during acts of war. This phenomenon is marked by many symptoms. The most prominent ones are startle reactions, fears, nightmares, fatigue, restlessness and irritability. Although combat exhaustion was only one of the war neuroses or 'gross stress reactions', other psychiatric problems that made fighting impossible for the soldier were not frequently mentioned. It is very likely that disorders other than combat exhaustion occurred infrequently during World War II.

It is important to be aware that symptoms were not confined to those who broke down. A strict division between patients and healthy soldiers is untenable. Most soldiers suffered from all kinds of stress reactions, sometimes even to a serious degree, but did not break down. This evokes the question of why, at a certain point, soldiers did suffer from combat exhaustion.

The debate about the causes of psychological breakdown in soldiers, which took place in the forties, is still relevant to current research. It is clear that disorders resulting from extreme experiences cannot be retraced to existing neurotic traits in the individual. An interactionist view (Endler & Magnusson, 1976), in which both situational and personal characteristics are accommodated, is more appropriate for understanding the disorders.

The terminology concerning the social and psychological aftereffects of combat has been particularly confusing. All kinds of terms were used simultaneously without having been defined and delineated in a concise or clear manner. The modern concept of 'posttraumatic stress disorder' appears a suitable umbrella concept for the various disorders.

It is striking that it was not until a considerable time after the Vietnam War had ended that attention was paid to the effects of this war. A considerable minority of the veterans suffer from problems such as nightmares, fears, recurring memories and physical complaints. Feelings of guilt occur frequently: guilt about losing comrades and about being alive while others have died. These symptoms appear to grow more serious in time. Explosive rage and violent impulsivity have often been reported in Vietnam combat veterans. Several veterans have a history of substance abuse and difficulties with authorities (Peterson et al., 1991). Significant problems in the areas of intimacy and sociability are commonly found.

Finally, a contemplative remark: by far the most findings in this chapter concern American, Israelian and West European soldiers. Information on German, East European and Vietnamese soldiers is lacking.

CHAPTER 4

DISASTERS

Disasters have a far-reaching influence on human existence. In the past, populations have been decimated by catastrophes such as floods, earthquakes and epidemics. During the great plague epidemic raging between 1347 and 1350, one third of the entire population of Western Europe died (Ziegler, 1976). Many disasters are accompanied by enormous material destruction. All that is considered to be secure is suddenly swept away. Total powerlessness is experienced, with tremendous chaos and desperation as the result.

In this chapter the focus is upon research in the social sciences on human behavior and experiences before, during and after disasters. Conclusions that apply to other traumatic situations may be drawn from this research field, since a disaster is probably the purest example of an overwhelming event that produces an intense feeling of powerlessness and disruption.

4.1 Introduction

As is the case with many important terms in the social sciences, there are many, often divergent definitions of disaster. Everyone immediately knows what the word refers to, but formulating an unambiguous definition is difficult. Five aspects of disastrous events emerge from the various definitions:

1. the powerlessness of the individual;
2. the loss of structure of daily life;
3. the large degree of destruction and human suffering;

4. the collective nature;
5. the sudden occurrence.

The first three aspects correspond with the characteristics of a traumatic experience, as described in chapter 1. Powerlessness refers to lack of control due to the overwhelming nature of a disaster. Loss of structure implies that existing securities – the anchors of daily existence – have been lost. Not until long after a disaster has taken place does life return to normal, and, in some cases, it never does.

A disaster is regarded as a collective stress situation. Many people are affected at the same time, so that damage does not only occur for the individual, but also for society. This collective nature of a disaster implies that the individual problems following a catastrophe should not be viewed separately from the social and cultural context. We will return to this later. Finally, the sudden nature of a disaster is emphasized in the various definitions. A disaster is a quick and dramatic event. It is an acute crisis, one that drastically disrupts daily life – in one sudden powerful moment.

Examples of disasters include tornados, tidal waves, earthquakes, floods, and mining and ecological disasters. A generally accepted classification of disasters does not exist. There have been relatively few attempts to develop a taxonomy. One of these attempts was made by Berren, Beigel and Ghertner (1980). This attempt at classification is enlightening in the sense that it points out important aspects of disasters. However, it is quite abstract and barely elaborated. Moreover, classifications have not been used in research and therefore we will not discuss them further.

A difference is sometimes made between natural disasters and man-made disasters, but this distinction is not clear. Many natural disasters have partly been brought about by human activities. For example, a flood may have been caused by a neglect of dikes and dams. Aside from that element, catastrophes such as a mining disaster or a devastating explosion (sometimes mentioned as technical disasters), often do not differ in appearance from a 'natural' earthquake or flood, during and after their occurrence. However, during recent years a new type of man-made disaster has emerged. Ecological or technological catastrophes such as the nuclear plant disasters in Chernobyl (USSR) and Harrisburg (USA) have a large impact on mental and physical health.

A historical overview

One of the first observations of a disaster by a psychologist was made by William James (1911). After the San Francisco earthquake of 1906, he was surprised by the high spirits and stability of the victims, and by the speed with which people created order from the chaos. The Swiss physician Eduard Stierlin (1909) was probably the first researcher to systematically and extensively observe the reactions of survivors of disasters, such as the earthquake of Messina (1909) and the large mining disaster of Courrières (1906; more than 1000 laborers killed). Stierlin mainly focused on the acute reactions to these catastrophes, but he also described several long-term symptoms such as sleep disturbances, anxieties related to specific aspects of the original event, and feelings of tension and depression. He mentioned the relatively calm and apathetic behavior of survivors directly after the disaster, sometimes combined with euphoria. This last phenomenon represented, in Stierlin's opinion, masking of repressed anxiety and fright. Stierlin's unique studies were, however, largely overlooked in traumatic stress research.

In the period between the world wars, research on disasters was conducted sporadically. After World War II, the systematic study of disasters was initiated within the social sciences research in several countries. Vroom (1942) examined psychological and psychiatric symptoms in the population of the Dutch town Den Helder during and after aerial bombardments. According to Chapman (1962), federal authorities provided the impetus in the United States in an effort to gain more insight into the implications for human behavior resulting from the nuclear arms' destruction. Thus, the effects of the massive bombing of Japanese and German cities were studied by a research team. Other examples of these first studies include: Titmuss' research (1950) on the problems surrounding the evacuation of large segments of the London population during the air war above this city and Tyhurst's study (1951) of the different phases in human behavior during disasters. In 1952, a special research committee to study human behavior during and after natural disasters was founded in the United States.

The study of disasters flourished in the beginning of the sixties. During this period, mainly sociologists concerned themselves with the field (Baker & Chapman, 1962: Barton, 1969; Haas & Drabek, 1970). The interest appears to have diminished somewhat thereafter. Nevertheless, there were disaster studies conducted in the seventies. For example, the various studies on the effects of the Buffalo Creek flood, a disaster in West Virginia, are important in this context. The attention increased during the eighties with the growth of the field of traumatic stress.

Disaster research has been conducted both in the United States and in other countries. The early work of the Swiss researcher Stierlin has already been mentioned. In West Germany, Ploeger (1974) analyzed the survivors of a mining disaster near Lengede. In Nicaragua and Peru, several researchers studied the effects of severe earthquakes (Cohen, 1976; Janney, Masuda & Holmes, 1977). In the Netherlands, the effects of the sudden and severe flood of 1953 were studied (Ellemers, 1956). There is an impressive and growing body of knowledge on disasters in Norway (Weisaeth, 1989a, 1989b) and Australia (McFarlane, 1989; Raphael, 1986).

A general impression

Social scientific research on disasters makes a rather chaotic impression. Reviews are still rare (exceptions include Kinston & Rosser, 1974 and Cohen & Ahearn, 1980).

Research is influenced to a great extent by the researcher's background, thus it is difficult to integrate the various research findings. A marked difference exists between the interpretation of human behavior by sociologists and by psychologists and psychiatrists. The former are primarily interested in the phenomena of social disorganization. They describe the immediate reactions to disasters in a neutral way. Some researchers (Dynes & Quarantelli, 1976; Taylor, 1977) strongly emphasize the active behavior of people following a disaster and tend to trivialize emotional problems. Psychoanalytically oriented researchers write more about various defense mechanisms such as avoidance, and they sometimes tend to focus on fear and denial to such a degree that it would appear as though everyone were numb, inactive and bewildered following a disaster (see, for example, Greenson & Mintz, 1971).

Few studies are systematic from a methodological point of view. Quantitative data have been lacking, but in the last ten years more quantitative studies have been conducted. Still, long-term effects are hardly studied, and control groups are usually absent.

In order to analyze human behavior during catastrophes, a classification has been made of the different phases of a disaster. Each phase is supposedly accompanied by specific experiences and behavior. A classic classification is the one developed by Powell and Rayner (1952), which consists of seven phases. This classification is too precise to fit reality. The separate phases are often difficult to distinguish. Nevertheless, some classification is useful in clustering the different phenomena surrounding disasters. Although such distinctions are sometimes arbitrary, they may, from a heuristic point of view, contribute toward understanding.

To that end, we reduce the classification of Powell and Rayner to a simpler version, which resembles those made by Tyhurst (1951) and Kinston and Rosser (1974). We distinguish between:

1. the phase before the disaster (anticipation phase);
2. occurrence and immediate effects (impact phase);
3. short-term effects (during the first few months after the disaster);
4. long-term effects.

Each of these phases will be described on the basis of findings from disaster research.

4.2 Before a disaster

Some disasters occur without any warning. The occurrence of others is known hours in advance. For still others, the threat has been present for years. However, the impact is nearly always sudden.

When threatened, people search for security. Something that they can barely understand or comprehend threatens to take place. They attempt to grasp the meaning of the alarm. The less clear the situation and the more contradictory the announcements concerning the impending disaster are, the greater the confusion and the fear. On the other hand, however, the warning of impending disaster gives people the opportunity to prepare for danger, not only materially, but also psychologically. They can prepare for the worst. It has been established that those who have had little or no opportunity to prepare themselves for a shock, experience more problems later and are less capable of adjusting to the new situation than those who saw the event coming (chapter 9).

Several authors emphatically state that only a minority of the people involved overreact to an impending catastrophe. Few people deny the existence of the danger in an absolute way, and few react in an exceedingly emotional manner (Chapman, 1962; Weiss & Payson, 1967). In her monograph on disasters, Wolfenstein (1957) concludes that these persons already possess an unstable personality, so that coping with the impending threat becomes troublesome for them.

Even when given some warning, most people employ a certain denial of the potential disaster. They refuse to realize that a catastrophe could happen to them. They have a vague sense of personal invulnerability. Wolfenstein

(1957) speaks of mythical thinking. Moreover, responsibility is often shifted to the authorities, who are supposed to possess the knowledge and means of preventing the impending disaster. Denial, however, is used in varying degrees. There are examples of the tendency of people to trivialize or even ignore the warning. One extreme example is a serious flooding of the Rio Grande River in Texas and Mexico, where a large crowd cheered as it watched the rapidly rising water (Kinston & Rosser, 1974).

People also tend to ignore warnings because they fear others will think they are reacting foolishly. Many receive the warning from relatives, neighbors and friends. They then determine the truthfulness of the message by watching what others think and how they react, and by discussing it with them (Dynes & Quarantelli, 1976). The mass media may be very important in announcing danger, but this news is usually also heard in the presence of others. That is why the *context* rather than the *content* of a message, is important (Dynes & Quarantelli, 1976). Social comparison plays a prominent role in the period before the disaster.

Chapman (1962) points out that human behavior just before an overwhelming event is not only determined by the preservation of life. Other factors are also relevant. During the heavy bombing of London in World War II, many evacuees preferred returning to the dangerous city, as opposed to staying in the country where they were confronted with other ways of life. Nor do many people run away when a disaster is imminent.

4.3 During and immediately after the disaster

As previously mentioned in the introduction of this chapter, researchers distinguish between a number of phases in the phenomena during and immediately after the disaster. Frederick (1980) names three phases: 'initial impact', 'heroic' and 'honeymoon'. During the occurrence of the disaster, the effect of the stress factor is maximal, direct and inescapable. This initial impact phase usually lasts several minutes to one hour, although under specific circumstances it may last much longer. Subsequently, the heroic phase takes place, during which survivors, along with rescue teams, frantically search for relatives and friends. After approximately one week, the so-called honeymoon phase begins, in which the familiar pattern of life is gradually rebuilt with a great deal of support from others.

Although Frederick's division initially appears illuminating, the problem is that the phenomena observed cannot usually be reduced to one single phase. Furthermore, nearly every author uses a personal subdivision of the period during and immediately after the disaster.

In a study that has become a classic, Tyhurst (1951) distinguished three types of reactions to a catastrophe. His division, often mentioned in the literature, is clarifying, although the percentages suggest an exactness that has not been supported by research findings.

According to Tyhurst, there is a group of people (approximately 12-25%) who behave in a cool and collected manner and who immediately direct themselves to lessening the distress of others. There is also a large group (approximately 75%) who are numb and bewildered at first. These people behave more or less automatically. They fall back on what they have learned in the phase prior to the disaster or during earlier disasters. This state of confusion and numbness is also reflected by the difficulty experienced later in remembering exactly what happened (Cohen & Ahearn, 1980). For most people, the bewilderment does not last long. They soon become active and begin to search for relatives and friends, like the former group. Finally, a third group of people (approximately 10-25%) exhibits excessive, inadequate reactions. These individuals are completely confused and lose control of their movements. Both extreme aggression and severely stereotypical or apathetic behavior (freezing) may result. However, as already mentioned, these percentages are not supported by empirical evidence. The 'cool and collected' group in particular appears to be larger relative to the other two groups.

As pointed out, people attempt to gain insight into the events during the phase preceding a catastrophe. They try to interpret them from their existing frames of reference. One good example is provided by Friedman and Linn (1957), two psychiatrists who, as passengers on a French ship, witnessed the disaster of the Andrea Doria and the rescue operation of the passengers on this Italian cruise ship. One passenger, aboard the French ship which was performing the rescue operation, heard a noise at 2:00 A.M. He looked out the window and saw countless lifeboats carrying people wearing orange life jackets. Big lights were aimed at the boats. The man got the impression that there was a party going on, and went back to bed. Just before he fell asleep, he realized that all the activity on the water had to mean something else.

The interpretation of the disaster from existing frames of reference implies that people retrace the phenomena to events that have taken place before. In his monograph on the Dutch flood disaster of 1953, Ellemers (1956) illustrates this with the following example; at first glance, the water entering the homes was perceived as caused by a broken water pipe or a leaking faucet.

As in the period prior to the disaster, forms of mythical thinking or illusions may play a role in the first reactions to a catastrophe. Wolfenstein (1957) mentions the illusion of centrality. Someone believes the disaster was aimed at him and perhaps some people in his environment, and that they are the only victims. The victim tends to withdraw emotionally, and be concerned only with himself. Often the illusion of centrality is accompanied by the idea that the disaster is a punishment or curse for sins committed. This illusion may be the result of the previously mentioned concept of personal invulnerability that a person, in spite of the real threat, possesses prior to the disaster.

The victim quickly discovers that his surroundings are not any less damaged than those of others, and that his first impression is inaccurate. This produces a great shock. According to Wolfenstein, the individual feels abandoned; he is not the only victim any more. He may feel that all means of help and support have disappeared or decreased. While a person is alone, the feeling of having been abandoned is very strong. When a person is in the company of others after the disaster, the feeling is less noticeable.

The often heard assumption that panic occurs during a disaster, is hardly correct. Panic only occurs under very specific circumstances, namely if:

1. people perceive an immediate and serious danger to themselves;
2. they feel that there is only one means of escape, which threatens to become blocked;
3. they receive contradictory information about 1. and 2. (Janis, Chapman, Gillin & Spiegel, 1955).

Examples of the above are the outbreak of fire in a cinema or riots in a football stadium. Even then, not everybody panics. In general, people behave quite calmly, rationally and bravely (Chapman, 1962; Cohen & Ahearn, 1980).

Weisaeth (1989a) studied the impact of a tremendous explosion in a paint factory in Norway. He was able to interview all 125 employees present at work when the explosion occurred (this response rate of 100% is unprecedented in traumatic stress research). Weisaeth found that most individuals exhibited adaptive behavior in the disaster situation, although about 20% carried out some degree of inadaptive behavior such as inhibited reactions.

The literature on disasters confronts us with a difference of opinion about the psychological symptoms that are said to occur immediately following a disaster. Many sociologists (Fritz, Frederick, Quarantelli) emphasize the responsible and enterprising behavior of those who are affected by a disaster. They argue that disasters have both positive and negative effects, and that there is little increase in mental illness. On the other hand, psychiatrists pay more attention to the bewilderment and helplessness evident immediately after the disaster.

It is, however, not a question of one or the other phenomenon. Both forms of behavior occur during and immediately after a disaster and there is a sequence of types of behavior. The initially numb and bewildered feeling which, according to Tyhurst, is accompanied by indecisiveness, is soon replaced by active and situation-oriented behavior. The survivors attempt to comprehend all that has happened. Friedman & Linn (1957) described the victims of the Andrea Doria disaster as initially passive, slow and docile. It was as though they were numb. This condition soon disappeared and everybody began to talk excitedly and with great emotion about what had happened. How rapidly this change takes place depends on the circumstances. If the catastrophe continues, the numb feeling may continue to exist as well, as Cohen (1976) reported about the earthquake in Nicaragua.

Moreover, the symptoms of numbness and activity may occur simultaneously. A person reacts emotionally numbed, but at the same time takes the appropriate action to save his skin.

'Honeymoon phase'

The survivors soon begin to search for relatives, friends and acquaintances. This search for survivors is a very dominant activity. Fritz and Marks (1954) found that less than half an hour after a tornado hit Arkansas (U.S.A.), 78% of the survivors were searching for missing persons or were involved in other forms of giving aid.

During the phase following the disaster, people characteristically need company and need to express their feelings about the disaster and the losses incurred. Differences in social status disappear temporarily as a result of the overwhelming catastrophe. Everybody is focused on saving their own lives and possessions as well as those of others. Because social barriers are absent, a strong sense of community emerges. For this reason, the previously mentioned term 'honeymoon phase' is used as a label for the period beginning approximately one week after the disaster and sometimes lasting several months.

Stress reactions and health problems

What types of stress reactions and health problems may occur after a disaster? After the feeling of bewilderment described above has disappeared and the search for relatives and friends has begun, the feelings resulting from the disaster surface. Fear is by far the most prevalent reaction. Aside from fear, sleep problems and startle reactions occur frequently, as do feelings of restlessness and tension. The literature (Cohen & Ahearn, 1980; Titchener,

Kapp & Winget, 1976) lists the following as short-term reactions: gastrointestinal complaints, memory and thought disturbances, nightmares, and disruption of daily activities. Hallucinations and delusions occur incidentally in serious cases. Anger and irritation are rare at first, but occur more frequently later on, as will be shown in the following section (Cohen & Ahearn, 1980; see also Friedman & Linn, 1957).

During the initial period after the disaster, the event is reexperienced time and time again. Memories suddenly emerge. Situations similar to the original traumatic event provoke the same emotions. People talk about the event continually. 'It is this reliving again and again, that acts to gradually extinguish, through repeated exposure in imagination, the fearful stimulus' (Wolfenstein, 1957, p. 135). This reliving (intrusion) has been discussed in chapter 2 and will be looked at from the perspective of a coping theory in chapter 8.

In some cases, however, total repression takes place. The survivor is not able to talk about the experience, and does not want to. Wolfenstein remarks that the need to reexperience in memory is proportional to the lack of preparation before the disaster. Adequate anticipation provides the opportunity to reduce tensions later on (chapter 9).

We have already noted that people become active soon after a disaster. Emotional reactions seem to recede into the background after a few days (Chapman, 1962). However, not everyone recovers this rapidly. One group of survivors exhibits the so-called 'disaster syndrome' (Wallace, 1956). They remain passive and withdrawn. Emotions and reactions to new events fail to surface. This pattern may continue for several hours and constitutes a very strong degree of avoidance.

It is estimated that 10% of victims need some kind of psychological aid immediately after a disaster (Kinston & Rosser, 1974). However, the percentage varies with the characteristics of the disaster, such as material damage, the number of wounded and the sudden nature of the event.

Health problems are also expressed in the official illness statistics. On June 24, 1972 an area in Pennsylvania was struck by a heavy flood that caused half a billion dollars worth of damages. Zusman (1976) studied the figures of the health center in one of the affected cities in the first two months following the disaster compared with the same period the previous year. The total number of patients increased 21%. In particular, there was a 28% increase in patients diagnosed as suffering from cognitive and affective disorders. However, Zusman also found a remarkable 64% decrease in the number of patients diagnosed as suffering from disorders in social relations. It is likely that this finding indicates a 'honeymoon phase'.

4.4 Some time after the disaster

Some time after the disaster, assistance and rescue actions are gradually reduced; bureaucratic problems start slowing down all types of activities and social distinctions reappear. The 'honeymoon phase' disappears and an atmosphere of disappointment and disillusion emerges. Survivors increasingly come to realize that many of their acquaintances are gone forever, which adds to these negative feelings.

According to Frederick (1980), this disappointment and disillusionment is expressed in anger and resentment towards the authorities, who do not provide the expected support. These hostile feelings toward authorities are common among survivors of disasters and are often mentioned in studies of victims of violence (chapter 5). Victims are suspicious of neighbors and acquaintances who appear to be better off. The sense of community declines. However, the survivors eventually realize that they themselves must rebuild their lives and environment. The bitterness, which may last up to approximately one year after the disaster, is then replaced by the reorganization and reconstruction of daily life.

The prolonged effects of catastrophes have received little attention in disaster research, although there has been some discussion in the literature. According to several authors (Chapman, 1962; Dynes & Quarantelli, 1976), the above-mentioned expressions of emotionality which immediately follow a disaster, rarely result in serious long-term psychological disorders. On the other hand, the impression has sometimes been, and continues to be (especially in the mass media), that a disaster may cause dramatic effects, such as mass psychosis and a substantial increase in the number of suicides, even after a lengthy period of time. This is not correct. In some cases people even feel that their lives improved after a disaster (Taylor, 1977). Nevertheless, the research conducted during the last 15 years has clearly shown that, in the long run, certain distressing psychological and social symptoms do occur as a result of a disaster.

One of the first studies in this field concerned the survivors of the Waco-San Angelo (U.S.A.) tornado in 1953 (Friedsam, 1962). After four months, 85% of the black population and 53% of the white population greatly feared unusual weather conditions. In a different study, Catanese (1978) interviewed eight survivors, residing in Southern California, about the Tenerife plane disaster, where a Dutch and an American jumbo jet collided. After five months, all these people exhibited characteristics of a traumatic neurosis, as described by Fenichel (1945).

The effects of a flood caused by heavy rains in the British town of Bristol were studied by Bennett (1970) in a particularly methodologically sound study. He compared the health care figures of the previous year with those the year after the flood. The study involved 770 people who resided on streets that had been flooded and those from streets that the water had not reached. The data were collected either through interviews or by using medical statistics. In the flooded area of the town, the number of consultations increased during the year after the flood. Further analysis showed that the frequency with which a doctor was consulted was related to the height of the water level in the house during the flood. The number of referrals to a hospital also increased for the group of victims; the number almost doubled, whereas the control group showed no changes. It is noteworthy that all of these findings apply more to men than to women. Also striking in Bennett's study was an increase of 50% in the mortality rate of the group of victims during the year after the flood. This increase occurred primarily in the age group ranging from 45 to 64. Cancer constituted the primary cause. The control group showed no changes in mortality.

After a heavy flood in Pennsylvania (U.S.A.), the mortality rate among victims also increased. In this case, deaths were primarily due to cardiovascular diseases, not only in the first few months (Zusman, 1976), but also during the three years after the catastrophe (McGee & Heffron, 1976). This would indicate that shock leads to a feeling of desperation and helplessness, which blocks adjustment to the new situation. It is sometimes referred to as the 'given-up syndrome' (McGee & Heffron, 1976).

Inspired by Bennett's research, a group of Australian researchers (Abrahams, Price, Whitlock & Williams, 1976) studied the effects of a severe flood in Brisbane (Australia). They reached largely the same conclusions as Bennett: visits to the doctor, psychological symptoms and the use of pills increased after the flood. They did not, however, find an increase in mortality.

In an analysis of the health statistics of a rural community struck by an eruption of the volcano Mount St. Helen (U.S.A.), it was also discovered that all types of illness statistics increased as compared to the situation before the disaster (sometimes drastically; over 200% for psychosomatic and psychological complaints), as did the mortality rate and various indicators of criminality, such as vandalism and fighting within the family. These changes occurred particularly during the first four months following the disaster (Adams & Adams, 1984).

Not only the mortality rate, but also the birth rate may change after a disaster. One year after the mining disaster in Aberfan (Wales), in which 116 children and 28 adults were killed by an avalanche of mining waste, the birth rate increased spectacularly. After five years, the birth rate had gradually decreased to the normal level for that area (Williams & Parkes, 1975).

This brings us to the potentially positive effects of a disaster. Barton (1969) remarks that people sometimes put so much energy into rebuilding their community and lessening each other's burdens, that the resulting living circumstances are better than the original situation. Aberfan, mentioned above, became a flourishing community several years after the disaster (Williams & Parkes, 1975). Authors such as Dynes and Quarantelli (1976) believe that the losses caused by a disaster are easier to bear than would normally be the case, because:

a. the loss is interpreted in terms of relative deprivation (people compare their losses with each other and conclude that they are reasonably well off) and;
b. the mourning process is stimulated by the increased social cohesion (the losses are discussed so frequently that coping with stress is optimal).

In the above discussion, aside from noting some medical statistics, attention has primarily been focused on the effects of a disaster within a one year period. However, there has been some research in which survivors were studied after a longer period (two to four years). This will be the topic of the next section.

4.5 Several years after a disaster

In 1972, a prolonged period of rainfall caused the collapse of a dam in a long and narrow valley in West Virginia (U.S.A.). The Buffalo Creek flood, in which 125 people were killed and damages amounted to 50 million dollars, has been of great importance to extreme stress research for several reasons:

1. this is one of the few disasters where the long-term effects have been analyzed;
2. several researchers have been involved;
3. extensive attention has been paid to the sociocultural background.

During the first 24 hours after the flood, 88% of the 4000 people who were forced to leave their houses, were tormented by bewildered and incoherent thinking. There was little crying and few other expressions of grief. Human awareness had been swept away by an 'overkill' of far-reaching events. Fear, guilt and grief were not present during these first 24 hours (Harshbarger, 1976).

In an earlier section, we pointed out that this psychological numbness immediately after a shock does not necessarily exclude activity. This is confirmed by the Buffalo Creek flood (Rangell, 1976). During the disaster, people behaved in a goal-directed and effective manner. Erikson (1976) very clearly describes how people behaved rationally in the emergency and how they managed to save themselves and others from the swirling water.

The disaster had an enormous effect upon the area. The majority of the houses had been destroyed. People did not return to their own homes, but went to special emergency accommodations. Nearly everyone had lost a relative or friend. The entire environment had changed; the color of the landscape had changed from green to the grey of the mud and debris. Some settlements had vanished completely. Erikson (1976) states that, in addition to the individual trauma, a collective trauma had occurred. The community, including all its social ties and contacts, had been destroyed.

Even more than two years after the disaster, the number of problems remained quite substantial. We will discuss the analyses by Titchener et al. (1976) and by Lifton and Olson (1976). The first interviews were conducted 14 months after the disaster, while the last took place at 28 months. It should be noted that the interviews were conducted exclusively with survivors who were involved in lawsuits for damages.

Titchener et al. found the following four groups of symptoms in 80% of the survivors:

1. continuous fear: phobias and obsession with water, rain and wind;
2. depression due to guilt, anger, resentment and the so-called 'death imprint' (see below);
3. increased smoking, drinking and a change in eating habits;
4. loss of interest in sexuality, friendships and favorite pastimes.

Lifton and Olson (1976) conducted 41 interviews with 22 survivors during various periods following the disaster (respectively 14, 22 and 27 months after the disaster). The effects they discovered are clearly similar to those from the study mentioned above. After 27 months, some or all symptoms of the so-called 'survivor syndrome' were present in all the survivors. This syndrome was originally formulated by Lifton, based on his study of survivors of Hiroshima. It will be discussed in a slightly modified version in the chapter on survivors of concentration camps (chapter 6). The five manifestations of the syndrome are:

1. 'Death imprint' and related fears. 'Death imprint' refers to memories and images of the disaster that are connected to death and destruction. The fear may be primarily expressed in nightmares, which occur regularly, even after two years.
2. Guilt about the death of others. The survivor experiences guilt because others have died while he is still alive. No matter how irrational it may

be, he feels guilty because he has failed to save them. This 'death guilt' also emerges strongly in dreams.

3. Psychological numbness, which is expressed in apathy, depression, withdrawn behavior and lack of interest in friendships and social activities. According to Lifton and Olson, this is perhaps the most universal reaction to a disaster. It is partly a continuation of the perplexing condition during the disaster, which, in fact, served as a defense against the overwhelming situation. On the other hand, this numbness concerns the need to avoid the various forms of guilt and fear. This psychological numbness is accompanied by various psychosomatic symptoms, such as gastrointestinal disturbances, fatigue and pains, and by manifestations of amnesia and passivity. This numbness is closely related to the defense mechanism of denial.
4. Deteriorating relationships. Survivors are often suspicious of one another, and exhibit their pent-up rage and anger against nearly everyone else.
5. The meaning of the disaster to the individual. Is the surviving individual able to find a significant meaning to the confrontation with death? Some victims of the Buffalo Creek flood found comfort and resignation in religion, but many experienced the disaster as man-made.

Sociocultural dimensions

Which causes can be distinguished for these long-term effects? Important factors in this context are the sudden occurrence of the disaster, the continual reminders of it (change in the landscape, the disappearance of houses) and the legal aftermath of the catastrophe.

In addition, the sociocultural nature of the area and its population is very important (Erikson, 1976). Buffalo Creek was a highly traditional community, which had been ruled by mining companies since the twenties (Hollander, 1970). Sometimes an entire region is owned by such a company. These companies are notorious in the United States for their tough measures against unions and control of the community.

The effects of the Buffalo Creek flood must be seen in terms of this sociocultural background. The survivors remained in the area, where they were confronted with memories of the disaster every day. Many continued to be employed at the mining company. The had few opportunities for other work or cultural or social activities. The inhabitants were put up in emergency accommodations, without consideration for existing relationships between neighbors and relatives, so that the community ties, which were essential to them, were severely disrupted. In short, the phase of reorientation and reconstruction was extremely difficult in this small, isolated community.

The Xenia tornado

The fact that a disaster does not always have to have the same effects becomes evident when we compare the Buffalo Creek flood with another well-documented disaster: the tornado that struck the small town of Xenia in Ohio (U.S.A.) in 1974, and resulted in 33 fatalities, 1200 wounded and much material damage.

Employees of the Disaster Research Center of the Ohio State University interviewed 350 survivors and conducted research with the aid of questionnaires after six months and after one year (Taylor, 1977). The effects after six months were not as serious as the researchers had expected. Few people experienced anxiety. However, 50% did feel more nervous, while the same percentage was found for depression. Furthermore, some experienced insomnia (27%) and headaches (25%). Additionally, an increase was found in the number of visits to the first-aid center, in the use of pills (valium, etc.) and in juvenile delinquency. The use of alcohol, however, had decreased.

After this disaster, positive effects also occurred. Many people felt that their experiences had indicated they could handle a crisis. Moreover, according to 27% of the inhabitants, social relationships had improved; only 2% stated they had deteriorated. Yet community life was not rapidly rebuilt. Shops remained closed and many people moved to different, far away areas. The price of land increased, as did other costs. People thus experienced problems in daily life, such as a lower income and problems concerning daily needs.

We must mention, however, that the researchers' perspective clearly played a part in the image they presented of both communities. Xenia was studied by researchers from Ohio, who tended to underestimate psychosocial problems. A publication on the tornado was entitled 'Good news about disaster' (Taylor, 1977). The researchers of Buffalo Creek, on the other hand, may have been somewhat too pessimistic. Their samples were based primarily on those who sued the mining company that was held responsible for the disaster, and who may therefore have suffered more.

Unfortunately, the survivors of the Xenia disaster were not again interviewed at a later date. It is therefore interesting to conclude this section with an outline of one of the few disasters studies in which the victims were interviewed several years after the event.

Follow-up studies of disaster

In the beginning of 1957, a tanker collided with a freighter near the East Coast of North America. The explosions on the tanker completely destroyed the ship and were responsible for the deaths of 10 people. Within two weeks, 27 of the 35 survivors were examined by two psychiatrists. After 3 1/2 to 4 1/2 years, they re-examined the survivors.

This disaster involved men between the ages of 18 and 55 with little formal education. Most of them had already spent many years at sea, and many of them had served at sea during World War II. They were therefore, in a certain sense, used to dangerous and threatening circumstances.

Immediately after the disaster, the survivors showed remarkably few physical problems, and were only slightly bothered by injuries. However, many psychological complaints were voiced and quite a few gastrointestinal disturbances occurred that could not be related to any physical causes. Four of the 35 men were admitted to a hospital for psychiatric reasons. The majority of the men, however, were in reasonable condition, despite certain complaints.

After 3 1/2 to 4 1/2 years it appeared that the psychological problems had clearly intensified. In the intervening years, 26 men (74%) had sought some type of psychiatric help. Additionally, an increase in depression, restlessness and phobic reactions was found. The survivors also reported several new complaints, such as hostility, distrust of colleagues and feeling isolated. The remaining physical problems primarily concerned back and neck pains.

Following the event in 1957, nearly all sailors returned to work within a few months. Four years later, however, is was established that 12 people had been forced to give up their work at sea, mostly due to psychological reasons. Others went out to sea sporadically, and by the four year follow-up only 12 of the 35 still regularly practiced their profession. All of those still working on ships stated they felt anxious and tense on board. This is striking in view of the fact that these were men with tremendous experience at sea, (who furthermore, earned far less on land).

On the basis of the complaints expressed by the survivors and of their own observations, Leopold and Dillon concluded that 24 of the survivors (71%) had deteriorated psychologically in the period between 1957 and 1960/1961. Seven had progressed, while three men exhibited no changes.

These negative changes clearly occurred more frequently in the somewhat older men. Education, marital status and war experience, on the other hand, were not related to the changes in time observed by the two psychiatrists.

Finally, we return to the Buffalo Creek disaster. Fourteen years after the flood a research team (Green, Lindy, Grace et al., 1990) returned to the small mining community in West Virginia to interview the adult plaintiffs that participated in the research conducted two years after the dam had collapsed (Titchener & Kapp, 1976). Only 39% of the 329 still living survivors completed the interview. The others had moved, refused, or could not be contacted.

The analysis of the changes between the 1974 and the 1986 examination showed a gradual decrease in psychological symptoms and an improvement in functioning with the passing of time after the catastrophe. In terms of the proportion of people who got better, worse, or did not change, 73% of the survivors improved on a symptom checklist, 23% remained the same, and 4% reported more symptoms in 1986. The improvements for women were more pronounced than those for men. Nevertheless, not all survivors were in the normal range of functioning so many years after the flood and the group as a whole showed significantly elevated levels of impairment relative to norms of standard instruments. Approximately 30% of the sample had clinically noteworthy levels of psychopathology 14 years after the disaster. Those who suffered from disorders in 1974 as well as in 1986 were different from the other survivors mainly with regard to more extreme and prolonged stressor experiences.

The researchers emphasized the social consequences. The disaster appeared to be a reference point in the history of the valley residents. The dam collapse was a shared experience. It marked the discontinuity in the community and landscape of the area. There were still anniversary ceremonies and the disaster had become a metaphor in the language and culture of the community.

Our conclusion is that disasters may lead to serious long-term effects. This is especially evident in the above-mentioned disasters, the explosion at sea as well as the Buffalo Creek dam collapse. In both cases the intensity and duration of the after-effects were primarily caused by the shocking nature of the catastrophe.

4.6 The aftermath of technological disasters

One of the ironies of the human condition is the appearance of new scourges once old ones have been conquered. A new type of disaster has emerged in the last decades: the technological catastrophe, sometimes called the ecological catastrophe, defined as the breakdown of man-made technical systems.

A nuclear reaction in Chernobyl (Soviet Union) exploded in 1986, resulting in the emission of large quantities of radioactive material. The soil of a vast area in the western parts of the Soviet Union became contaminated, but the extent of the area as well as the extent of the contamination were, and still are unclear. Acute radiation effects were only few, but late effects such as leukemia and other cancers may develop after years, even decades. The radioactive threat is invisible; there is an expectation of future illness, whereby the consequences are aggravated by the initial lack of information provided by the authorities and by the resulting ignorance of the population. Symptoms are: worries about one's own health, worries about the causal explanation with regard to any disease, worries about the future of the children, plans to move, distrust of authorities. Giel (1991) mentioned that the strong desire of the residents of the area for repeated measures of radioactivity and for physical examination of complaints of whatever nature showed that they were still striving to find out what had exactly happened. The main problem is the enduring threat of radioactive contamination.

An accident occurred at the Three Mile Island nuclear power plant near Harrisburg, Pennsylvania (U.S.A.), in March 1979. It involved the release of radioactive gas into the environment. Residents living near the Three Mile Island nuclear plant exhibited more symptoms of stress than a control group one year after the incident as well as at a 5-year follow-up. They showed significant deficits on measures of task performance and there were signs of chronic physiological arousal, such as elevated heart rate, increased blood pressure and higher levels of the hormones cortisol and norepinephrine (Davidson & Baum, 1986). They experienced more intrusive and avoidance thoughts than the control group. They were especially bothered by intrusive thoughts about the damaged nuclear reactor. This finding is relevant because Giel (1991) in his impression of the situation in Chernobyl did not expected intrusive posttraumatic stress reactions, such as startle reactions or flashbacks, as most people did not experience the catastrophe itself.

Remarkable are the findings on styles of coping with this technological disaster (Baum, Fleming & Singer, 1983). Residents reporting greater use of problem-oriented strategies of coping showed more symptoms of psychological and physiological disturbance than did residents reporting use of emotional management of the problems (e.g. avoiding the issue of the accident, working harder, putting it out of their minds). An explanation is that there is little that individuals can actually do to change the situation so that attempts to alter the situation may further diminish feelings of control and may result in frustration. One can better regulate one's emotional responses to the events that gaining control over a damaged power plant. Findings of another study (Cleary & Houts, 1984) also indicated that, contrary to traditional expectations about adaptation, people who took many protective actions, reported being active in organizations concerned with the incident, and sought out others to reduce their distress showed significantly more psychophysiological stress symptoms and were more upset. They developed a sense of hopelessness.

Characteristic for ecological or technological catastrophes are their delayed and chronic psychological effects. They hardly involve direct damage or injury to a population, but they result in long-term uncertainty about what may happen in the future, in particular because the severity and duration, and even the nature of the threat are ambiguous. The point of worst impact may not pass with the event. Furthermore, technological mishaps result in a loss of confidence, not only in future ability to control technology but also in the authorities.

4.7 Conclusions

Disaster research does not make a very consistent impression, probably because it has always occupied a marginal position in psychiatry, sociology and psychology. Little attention has been paid to the development of theory or systematic analyses.

Disasters are followed by various reactions. During and immediately afterward, people experience an enormous degree of powerlessness, which may be accompanied by emotions of fright, anxiety and desperation, although most people are initially bewildered and stunned. These reactions lessen rapidly, as most of those affected become actively involved in the rescue operations. Most people are reasonably capable of adjusting to the circumstances immediately after the disaster, although there is an accompanying increase in stress reactions and health problems. People who, for some reason, cannot become actively involved, will experience more and longer-lasting disorders (Abrahams et al., 1976; chapter 9).

Furthermore, it has become clear that disasters may lead to medium or long-term effects. Months, sometimes years later, the victims of serious catastrophes, such as a flood or an explosion, may suffer from effects, similar to those of other traumatic experiences: anxieties, sleeping problems, depression, irritability, loss of energy and somatic disturbances. The various symptoms fit well into the diagnosis of posttraumatic stress disorder, discussed in chapter 2. The characteristic dimensions of coping with severe stress, i.e. denial (in the form of the so-called psychological numbness) and intrusion (recurrent memories of the disaster), are evident.

It is important to emphasize that calamities have an impact not only on the victims and their relatives, but also on the rescue workers, such as police officers, fire fighters, physicians and nurses. Gruesome situations, body identification, time pressure, hazardous work environments and uncertainties in the worker's role may lead to a variety of somatic, social and emotional sequelae. Recently there has been a growing interest into the after-effects among disaster rescue workers and ways of assistance (Mitchell, 1986; Raphael, 1986).

Relatively little attention has been paid to the question of why one disaster results in more long-term effects than another. Yet there are some indications for this: the duration of the disaster, its unexpected nature, the social circumstances, the availability of aid and rescue and the loss of relatives and friends. Differences in the consequences of disasters are related to these aspects and not so much to the 'type' of disaster such as a tornado or a mining disaster. Moreover, we have shown that the sociocultural circumstances – expressed in both cultural norms and values, and in social support – are a very important determinant of reactions to natural disasters. The controversial issue in the combat stress literature about the relative weight of individual predispositions versus situational characteristics is rarely mentioned in disaster research.

Disaster research provides abundant information on the effects of extreme events, described with the aid of many, often fascinating, case studies. Relatively little attention, however, has been paid to two important questions:

- How does the individual process of coping with a disaster take place? How does an individual handle the problems?
- Which factors (such as the characteristics mentioned above) influence the occurrence and nature of the effects?

We will return to both of these questions later.

CHAPTER 5

VIOLENCE

Violence implies the infliction of physical suffering or damage, abuse of power, and violation of individual freedom. It is a serious threat to normal human existence. An individual is degraded to an object as a result of the violent acts of one or more other people.

There are many forms of violence: rape, assault, battery, torture, robbery, hostage situations, destruction, extortion and isolation. Nearly every form contains both psychological and physical acts of violence. Sometimes the threat of death is very explicit (e.g. in the case of a hostage situation); sometimes the violence is an intrusion into the pattern of daily life (e.g. a burglary). Various motives may spur the violence, such as financial gain, political convictions and/or feelings of displeasure. In this chapter we will limit ourselves to violence that occurs in times of peace.

5.1 Introduction

Crimes of violence occur frequently. In The Netherlands, a country with approximately 15 million people, 1,138,686 offenses came to the attention of the police in 1989. In Sweden (population: 8.5 million) this figure was 1,144,800 in the same year; in the former Federal Republic of West Germany (population: 61.5 million) 4,356,726 in 1989. The total number of offenses known to the police in the U.S.A. was 12,070,000 in 1983. These numbers have increased over the years; in 1970 265,732 crimes were reported to the police in The Netherlands.

Not every offense is known to the police. More accurate data with regard to criminality are determined independently from police registration, using the so-called victim-surveys: studies of large samples of randomly selected citizens. It appears that much of the vandalism, threats, acts of physical violence and theft is not reported to the police (Van Dijk & Steinmetz, 1979).

The experience of 'being a victim'

In a crime of violence, the individual is reduced to an object. He is denied freedom of action. Many victims of violence feel humiliated. Feelings of dehumanization are particularly present in a hostage situation, where one is an object of negotiation. Individuality and control of one's own existence are ignored, resulting in a severe degree of powerlessness. This is all the more distressing for the person involved, because his position as a victim is usually coincidental. The reason for his involvement in this hijacking or robbery escapes him: 'Why me of all people?'

Aside from a few reflective moments, an individual normally feels invulnerable. A crime of violence – or, in fact, any other serious life event – destroys this feeling. Suddenly one realizes that life cannot be taken for granted. A new aggressor could be waiting around every corner of the street. Janis (1971) refers to this as the phenomenon of injured invulnerability. Victimization seriously challenges the fundamental assumptions in one's day-to-day life (Janoff-Bulman, 1989): 'I never thought it could happen to me'.

A victim is suddenly confronted with a death threat and with the awareness of his own mortality. Van Dijk (1981) reports that people who were held hostage on a train for days, experienced a sense of loss: 'they have lost the naturalness to which they were attached'.

The central elements of a traumatic experiences described in chapter 1 – powerlessness, disruption and extreme discomfort – are clearly present in the case of a violent act. Added to this is the presence of an aggressor, who ignores the existence and integrity of the victim. In this chapter we will show that this factor is responsible for some important differences between this type of traumatic event and other serious life events.

The lack of attention

Until recently, the social sciences paid no attention to the psychological and social effects of violent crimes upon the victims. On the other hand, psychological and psychiatric studies of the offender had already been conducted during the early period of the modern social sciences. The victim was ignored by science and by society as well. It was not until the seventies that both academic and social interest began to emerge.

Attention to the psychological impact of violence has not only been stimulated by the expanding field of traumatic stress studies, but also by the victim assistance movement. In the late seventies, the realization grew that the judicial system was mainly directed at solving the crime and hardly at supporting the victim. The victim assistance movement has been influential in improving the legal position of the victim, in developing local support schemes for victims, and in improving police-victim interactions. These programs have been particularly successful in countries such as Great Britain and The Netherlands, where many local support agencies assist the victims of various types of crime. These organizations provide information on legal matters and compensation, they support victims emotionally and practically, and assist in other ways directly after the act of violence. However, because of the practical and often also ideological perspective of victim assistance, very few contributions to scientific research developed.

A sequential model

The effects of a violent crime may be meaningfully described in relation to the development over time of the reactions to the incident. A sequential model assumes that people will exhibit similar reactions following violent situations, and that these reactions will show a development over time. We must, however, remember that any sequential model is an ideal-typical formulation. Individual differences will be addressed later.

Several sequential models have been developed to describe the effects of violence upon the individual. The number of phases, the names and descriptions may vary, but the similarities are more relevant than the differences. We assume the following three phases:

1. disbelief and bewilderment;
2. reexperiencing, avoidance and emotions;
3. integration and recovery.

Figure 5.1 Phases in the reactions to violence

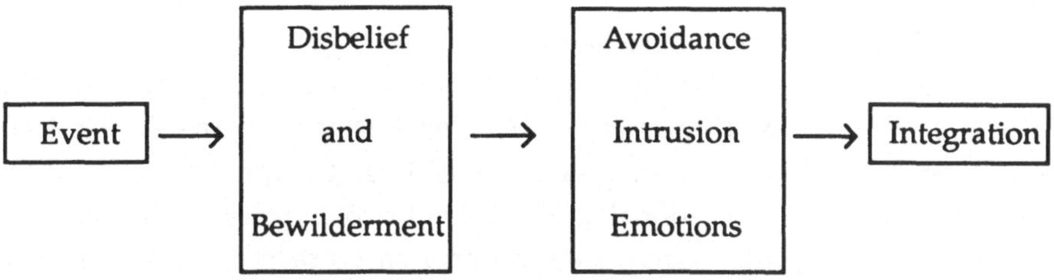

This division of phases integrates the various models found in the literature. Burgess and Holmstrom (1974), for instance, mention two phases for the effects of a rape, but their first phase contains an acute reaction and a period of physical and emotional reactions, which strongly resemble the first two phases in our division in terms of description and duration. Others (Bard & Sangrey, 1979; Sutherland & Scherl, 1970; Symonds, 1980) also distinguish three phases.

5.2 The first reactions: disbelief and bewilderment

The victim reacts to the sudden confrontation by an aggressor with disbelief and bewilderment. The victim can barely comprehend what is happening. Very characteristic of the first reaction to the event is disbelief: 'This can't be true'. As was the case in the disasters described in the previous chapter, the person is not yet fully aware of the reality. This phenomenon is indicated by terms such as numbness, disorientation and depersonalization.

The victim is confused and often experiences difficulty in talking about the event, particularly in the case of rape. The victim cannot remember details, and his thought processes are not clear or coherent (Bard & Sangrey, 1979).

The victim often feels helpless and abandoned immediately after the crime. This helplessness increases because of a need to function right away. The police must be informed, as must the insurance company. Some authors (Bard & Sangrey, 1979; Symonds, 1980) often point out the similarity of this helplessness to childlike behavior. The victim reverts to his childhood behaviors and feelings, when he may have reacted fearfully to the absence of his parents and was not capable of dealing with the situation.

Helplessness, disbelief and confusion are probably more characteristic than fear during the first moments after a crime of violence. Feelings of fear often emerge only after the threat has completely passed and the victim recalls the event.

Reactions of fear are more likely to emerge during the crime, when it is of a longer duration. During a rape, the victim is anxious, confused and terrified (Veronen & Kilpatrick, 1983). The hostages in the Dutch train hijackings reacted with a feeling of disbelief. Even though they realized the gravity of the situation rationally, this did not sink in emotionally. Simultaneously or shortly afterwards, victims experienced reactions such as shaking, a weak feeling in the knees, perspiring, crying and heart palpitations. Only afterward was the severity of the situation comprehensible, and this understanding was accompanied in some cases by extreme anxiety. However, this awareness of the situation often alternated with moments of feeling unreal and of estrangement, according to Van Dijk (1981).

A similar pattern of reactions during a violent event is mentioned by Symonds (1980). First the disbelief mentioned above emerges, followed by controlled action, which Symonds refers to as 'frozen fright', but which may also be defined as acute depersonalization. The victim is still emotionally numb, but he is capable of performing appropriate locomotive and cognitive functions. He performs the correct actions almost automatically. This reaction is functional in a prolonged situation such as a hijacking; emotions must be controlled.

Disbelief recurs after a long-lasting violent situation has come to an end. Immediately after their liberation, many hostages appeared dazed, distant and almost stunned (Van Dijk, 1981), probably in part because this particular train hijacking had been ended with a very forceful action by the authorities. It was difficult for the freed hostages to comprehend the change in their situation.

Burgess and Holmstrom (1974) interviewed 92 women who, over the course of one year, reported to a hospital in Boston because of rape. The women varied in age from 17 to 73 and originated from different social backgrounds. The women were interviewed immediately after their visit to the hospital as well as a considerable time later. A few hours after the rape, the researchers observed two styles of emotional reaction:

1. expressive: feelings of fear and anger were expressed by crying, restlessness and tension;
2. controlled: feelings were hidden behind calm, collected behavior.

The latter style is very similar to the 'frozen fright' phenomenon mentioned above. Van Dijk (1981) identified a similar dichotomy in the emotional reaction of victims immediately after the hijacking had ended.

5.3 Reexperiencing, avoidance and emotions

Following an act of violence, a victim must readjust to daily life. During this period the individual experiences a gamut of painful emotions, such as fear, anger, guilt and sadness. He is repeatedly occupied with the event and his moods alternate often, but the emotions and thoughts are frequently so unpleasant he tries to avoid them. This second phase in the effects of an act of violence starts several hours or days after the event, and lasts several weeks, although some victims may remain in this phase much longer. In this case, reexperiencing, avoidance and the alternation between these two are again emphasized. The defense against overwhelming emotions is regarded as a useful part of coping.

Diverse reactions are considered to be expressions of this defense. The victim engrosses himself in all types of activities; sometimes he is actively involved in solving the crime (Bard & Sangrey, 1979). Moving to another residence, going on trips and visiting relatives and acquaintances he otherwise barely saw, the victim may be hyperactive: another form of avoidance. In their research, Burgess and Holmstrom (1974) discovered that 44 out of the 92 raped women changed residences during the first year.

Another expression of defense is emotional inhibition, an emotional numbness that almost seems to be a continuation of the numbness from the first phase. This defense enables the victims to protect themselves from the painful emotions surrounding the event (Bard & Sangrey, 1979).

Sutherland and Scherl (1970) emphasize these symptoms to such an extent that they formulate a separate phase in the process of coping, which they refer to as 'outward adjustment'. After the initial shock has more or less disappeared, the raped woman returns to her normal existence. To the outside world it appears as if she has coped with the event. The woman is calm and pretends nothing exceptional has happened. However, according to both authors, this is only an apparent adjustment that is based on avoidance and rationalization of the fears, especially those concerning the aggressor.

All types of defensive phenomena occur following an act of violence, but the assumption of a separate phase of defense after a crime of violence does not seem correct in view of the fact that very few researchers distinguish such a period of external adjustment and internal defense. In his study of former hostages, Van Dijk (1981) established that there had not been a symptomless period between the hijacking and the appearance of symptoms or a breakdown.

Moments of defense and of reexperiencing tend to alternate. It is quite possible that, for some victims, one of the two poles is dominant for some time. In the extensive research of the hijackings in The Netherlands during the '70s, 168 former hostages who were interviewed – 59% of all hostages from those years – were questioned about their coping strategies. Only a minority (14%) reported they had avoided all things strongly associated with the hijacking, during the initial weeks. A larger number of the former hostages (40%) had, on the contrary, been occupied with it to a great extent, whereas the largest group (46%) sometimes avoided thinking about the hijacking and was sometimes preoccupied with it (Bastiaans et al., 1979).

A complete denial of recollections and emotions does not occur frequently, thus the assumption of a separate phase of defense is not justified. Partial defense occurs much more frequently. Between defensive moments, the victim focuses himself upon the event. The victim continually checks his recollections to judge where things went wrong and what could have been prevented. The event returns, both in countless conversations and in dreams (Bard & Sangrey, 1979; Sutherland & Scherl, 1970).

Defense

In the section above, the term 'defense' has often been used. Clarification is useful, because the term refers to symptoms that occupy a central position in the effects of serious life events.

Defense mechanisms, in the psychoanalytical sense, concern the techniques that the ego employs to protect itself from conflicts and unpleasant emotions. Every psychological manual mentions different lists of defense mechanisms. Characteristic of all defenses is that they:

a. reduce negative states (especially anxiety);
b. entail a certain distortion of reality;
c. are unconscious (Holmes, 1984).

With regard to the effects of violent crimes, the following defense mechanisms are especially common: repression (selective forgetting of conflicting material), suppression (avoiding stress-evoking thoughts), denial (paying no attention to the threatening aspects of a situation and re-interpreting them in a less unpleasant way) and rationalization.

The objection to a classification of defense mechanisms is that they cannot be strictly distinguished from one another. Brenner (1981) even regards postulating separate defense mechanisms as misleading. Instead, these are forms of mental functioning that may, under certain circumstances, have a defensive function. A person may behave defensively in all types of ways (fantasizing, denial, shifting the focus of attention), but this same behavior may have a different purpose in a different situation, such as satisfying needs. According to Brenner, there are no separate defense mechanisms. The central element of all defensive phenomena is that there is always denial: 'in defense, a person's ego is saying 'no' to whatever is the target of defense' (Brenner, 1981, p. 565).

For this reason, the terms defense and denial are used simultaneously and interchangeably in this study. This is in line with Horowitz (1976), who also uses the term denial, not in the sense of a specific defense mechanism, but in the sense of all the types of defensive strategies an individual employs.

Symptoms

The second phase in the reaction to traumatic events is characterized by the various symptoms in which intrusion and denial are expressed. Long lists of divergent reactions have been published (for instance, Holmes & Lawrence, 1983). We will discuss only the most relevant here. As an example, we will present a list from the Dutch research on large scale hostage takings (two train hijackings, two plane hijackings and four occupations of buildings). In these studies (Bastiaans, Jaspers, Van der Ploeg, Van den Berg-Schaap & Van den Berg, 1979; Jaspers, 1980; Van der Ploeg & Kleijn, 1989), the short-term and long-term consequences were investigated by means of interviews and questionnaires. The data are summarized in table 5.1.

Anxiety – in various types of manifestations – is one of the most important emotions, if not the most important. Anxiety is not always expressed during the first phase, but while reexperiencing the event, the individual suffers from real fears. These are clearly linked to the situation of the shocking event. Dutch research on former hostages, in which 26% of those interviewed experienced anxiety as the most important effect, indicated that in the first weeks following the hijacking – usually a train hijacking – victims were afraid to travel by train; loud noises such as airplanes frightened them terribly and they were afraid of being alone (Bastiaans et al., 1979).

The situational nature of the anxieties is also expressed in the symptoms following a rape (Peters, 1975). The next observations make this clear: A woman who had been raped in her bedroom was afraid to be alone at home; a woman who had been assaulted outdoors, was afraid to go out into the streets by herself (Burgess & Holmstrom, 1974).

Table 5.1 Psychological effects for those held hostage in The Netherlands during 1974-1978 (1)

Effects 2	short-term	long-term
Tension	41	—
Anxieties	34	29
Sleep difficulties	32	10
Preoccupation	28	10
Physical complaints	20	12
Concentration disorders	12	8
Irritability	8	12
Living in a daze	8	—
Depression	8	7
General feeling of insecurity	—	10
Feeling misunderstood	—	10
Aggression	—	7
Relativization	24	25
Enjoying more consciously	12	4
More intensive contact	6	10
Enhancement of self image	2	7

1. Percentages indicate the number of times a reaction was mentioned by the former hostages as one of the three (in the long-term 2) most important effects (N= 163). Short-term effects refer to the consequences in the first four weeks after the events; long-term refers to approximately two years later.

2. This table was set up by the authors on the basis of the data by Jaspers (1980); the sign — indicates unknown.

Principles based on learning theory (Defares, 1990; Ramsay, 1982) have provided insight into the development of such phobias. The pain and threat that emanate from the aggressor are unconditioned stimuli that evoke anxiety. The characteristics of the original situation – e.g. darkness, being alone, strangers – are associated with the unconditioned stimuli by means of classic conditioning, and thereby develop into conditioned stimuli. The result is that they evoke anxiety, even when the aggressor is absent. Stimulus generalization may occur. Someone who has learned to react to a situation in a certain way will also 'behave in this manner in similar circumstances'.

The individual subsequently learns to avoid the anxiety-evoking situation through operant conditioning (Holmes & Lawrence, 1983). The reaction of withdrawal reduces the anxiety and is therefore satisfactory. This tendency to avoid all types of situations develops with more frequency. Ultimately, this results in a phobia. By avoiding locations and persons connected to the original event, the person is protected from additional suffering. The phobia has a magical nature. This analysis, based on learning theory, is focused primarily on anxiety, and not on other reactions to violence. Despite this limitation, it has proven its value, especially in the psychotherapeutic treatment of disorders after traumatic experiences (chapter 11).

Occupying a special position among the reactions to a violent crime is *anger*. Some authors (Bard & Sangrey, 1979; Symonds, 1980) regard anger as a very common reaction, others barely mention it at all (Frank, Turner & Duffy, 1979). The victim has felt humiliated, has been powerless against the aggressor, but has not been able to express his anger, not even afterwards, because he usually no longer sees the offender. There is almost no direct way to vent the anger.

Anger is expressed in fantasies and dreams. At night the victim takes revenge on the offender, often in such a way that the dream itself evokes fears. The anger is also ventilated toward outsiders – often toward immediate relatives or the social worker – or toward himself. In the many instances of intrusion the search for an explanation for the crime is also pursued. 'If I had not done X, Y would not have happened'. Because it occupies an important position in the coping process, we will discuss this so-called attributional or explanatory process more extensively in chapter 8. It may be regarded as an effort to get a grip on the situation and is expressed in the most frequently occurring dream after a violent crime: 'the replica-dream'. The event is reexperienced, but with minor changes; the aggressive nature of the offender may be strengthened, for example, so that the victim may reduce his own guilt (Symonds, 1980).

This brings us to the *feelings of guilt and shame*, which are especially characteristic in women who have been raped. In an American study, the victims were asked about their feelings shortly after the rape. Although fear, anxiety and helplessness were dominant, having been mentioned by over two thirds of the women, feelings of shame and guilt were reported by approximately one third of the women (Queen's Bench Foundation, 1975). They reproached themselves for not having done more against their aggressor and felt guilty about the rape. The idea that a person is responsible for her own actions certainly played a part, as did traditional ideas concerning women. 'My father always said whatever a man did to a woman, she provoked it' (Burgess & Holmstrom, 1974, 983).

Depressive mood is also one of the reactions to a violent crime. The victim feels despondent and behaves passively. If these complaints are accompanied by eating and sleep disturbances and worrying and avoidance of social contacts, the victim may suffer from depression, a characteristic result of helplessness (chapter 8). Frank et al. (1979) studied 34 women who had reported to a regional center for victims of violence. Fifteen women were diagnosed, employing a standard test, as being moderately to very depressed. The most prominent symptoms were a despondent mood, feelings of guilt, loss of interest and reduced concentration. Eight of these 15 individuals were diagnosed by an independent psychiatrist as suffering from clinical depression.

However, the term 'depression' creates confusion, because it is used to indicate different disturbances, ranging from apathy and dejection to serious mood disorders (chapter 2). We therefore prefer the term depressivity or despondency.

Negative effects also include *physical complaints*, such as muscle tension, gastrointestinal complaints, general fatigue, sleep problems and, of course, the physical trauma resulting from the crime itself (Burgess & Holmstrom, 1974). More than one third of the Dutch former hostages felt they had deteriorated physically (Bastiaans et al., 1979).

After such a lengthy discussion of unpleasant effects, the question arises as to whether victims might also perhaps experience *positive reactions* afterwards. The Dutch research on former hostages provides an answer. After anxiety, tension and sleep problems, the most frequently reported short-term effect was: increased ability to put things into perspective (mentioned by 24%). In light of the hijacking, all types of matters that were previously regarded as important, obtained a certain relativity. For example, people handled money more easily. Also, victims mentioned the feeling of a more conscious enjoyment of life. People saw life through new eyes and experienced relationships more intensely (Van Dijk, 1981). These positive feelings occurred more frequently the longer the hijacking had lasted but were not really strong (Jaspers, 1980). In view of the rather unique nature of hijackings, in which the victims receive a relatively high degree of attention from the immediate environment and the authorities, it is not clear whether positive reactions also occur after other violent crimes. However, Veronen and Kilpatrick (1983) report somewhat disconcertedly that positive reactions also occurred after a rape in one or two cases.

5.4 The phase of integration

The intense emotions surrounding the event gradually disappear as the victims are once again integrated in daily existence. They talk and think less frequently about the event and, when it is brought up, are less upset by it. This third phase is described by terms such as integration, reorganization and recovery.

The theoretical considerations which Sutherland and Scherl (1970) formulated in their study of a small group of women who had been raped, may provide insight into the recovery process. In the last phase of recovery, two issues must be solved:

a. *The integration of the experience in the image of one's self and the world.* A raped woman will accept that the rape has taken place. The more or less natural, matter of fact character of the daily activities and expectations once again returns albeit in a different, somewhat more realistic manifestation. One's own vulnerability is not, or in any case much less frequently, dwelled upon. The question 'Why me?' will be asked less often. The person will also gain a better understanding of her own part in the crime, while, in addition, the negative feelings she has towards herself diminish.
b. *Coping with the feelings toward the aggressor.* The attitude of 'understanding' the aggressor's problems, which appeared in the so-called apparent adjustment in the previous phase, is replaced by anger about his abuse and her own lack of resistance. In terms of attribution theory (chapter 8), one might state that internal attribution – searching for an explanation for the crime within oneself, which is manifested in feelings of guilt and self-reproach – shifts to the background more during this phase of coping.

Although the life of the victim of violence 'calms down' after some time has passed, this does not mean that complaints and symptoms have disappeared. The studies of the victims of the Dutch hijacking actions (Van Dijk, 1981; Jaspers, 1980) established that roughly two thirds of them suffered from one or more effects some six months to one year later. After several months, the victims complained of anxieties: 28% of the people who were interviewed reported suffering from them. Also included in the long-term effects were the feeling of being misunderstood and a general feeling of insecurity. The former reaction was thought to result from people in the immediate environment growing tired of repeatedly hearing about the adventures of the hostage, while the person concerned still felt the need to talk about the event. The uniqueness disappeared, and the individual felt abandoned: 'aren't you over it yet?' A follow-up study (Van der Ploeg & Kleijn, 1989) showed that the amount of negative after-effects had decreased. However, even in the long run (6 to 9 years after the event) hostages still report anxieties, tenseness, sleeping problems, and psychosomatic symptoms.

A rather permanent and general feeling of insecurity occurred also in women who had been raped. In an American study (Queen's Bench Foundation, 1975), 89% of the 55 women who were interviewed reported that, in the long run, their sense of security had been violated the most. For long after the rape, they felt frightened in the dark or while alone and they behaved distrustfully towards strangers. Other reactions were the loss of self-esteem, the breaking up of social relationships and the disruption of sexual activities. This last effect is mentioned frequently in the literature (Veronen & Kilpatrick, 1983). What should be pleasant has become a continual reminder of violence.

The study of Burgess and Holmstrom (1974) showed the following as the most prominent long-term effects: increase in all kinds of activities of an avoidance-oriented nature (moving house, changing one's telephone number, etc.), nightmares and the previously mentioned anxieties and phobias.

These anxieties and phobias occupy a central position in the methodologically sound research conducted by Veronen and Kilpatrick (1983). As one of the few research projects on violent crimes, their study utilized control groups of comparable women who had not been raped. They compared both groups, each consisting of 20 women, several times after the event. After three to four weeks, the raped women showed considerably more reactions such as fear, dejection, low self-esteem and various expressions of discomfort. Three months after the rape, most differences had vanished, including those related to the daily functioning; however, the anxieties were still clearly present. Even after 6 and 12 months, these differences in anxiety (of men, strangers, the dark and being alone) between both groups, remained.

The question once again, is how many victims experienced these complaints? In the studies of Veronen and Kilpatrick, it was established that 17-25% had no complaints after one year. In other studies on raped women, percentages of 16% (Burgess & Holmstrom, 1974) and 27% (Peters, 1975) are reported. One third of the Dutch former hostages did not suffer any negative effects after one month (Jaspers, 1980).

Although reactions to the event are present in a substantial majority, this does not say how serious they are. In Jaspers' study, the former hostages were asked to indicate the gravity of the long-term negative effects; 43% reported these were few or absent, 30% regarded them as medium and 27%

called the reactions severe. It is worthy of note that Burgess and Holmstrom reached practically the same figures in their study of women who had been raped, even though they used a different method, namely that of observation. Although individual differences will be addressed later (chapter 9), we see here that the intensity of effects in the hijacking research were related to the duration of the hijacking and to certain personality traits (Jaspers, 1980). The primary high-risk group consisted of young women with a primary school education who were held hostage for a long period of time; this was also the case for the short-term effects.

Several other findings from the hijacking research are worth noticing. There is a clear connection between the short-term effects and the long-term effects. The stronger the one, the stronger the other will also be. If a person exhibits only a few serious reactions to the event during the first weeks, then their severity usually does not increase in the long term. Strong reactions shortly after the hijacking will not simply have vanished some time later (Van der Ploeg & Kleijn, 1989), while serious complaints do not, in the long run, emerge out of nowhere.

In the long run, the intensity of the emotions will diminish. In the previously mentioned research about women who have been raped (Veronen & Kilpatrick, 1983), much of the improvement occurred between the first and third months. The length of time over which fears and feelings of insecurity remain is not certain, as information is lacking. Although a follow-up study of raped women (Burgess & Holmstrom, 1978) mentioned that five years after the event not everyone had recovered, there were methodological objections to the measurement. Undoubtedly the memory will reemerge when the victim hears about a similar violent crime in his immediate environment. Nearly half of the Dutch former hostages declared that the fears, sleeping problems and other reactions reemerged for a short period when they heard about a new hijacking (Jaspers, 1980).

5.5 Methodological problems

Unfortunately, quite a few methodological objections pertain to studies concerning the effects of crimes of violence. It is important to examine the most prominent shortcomings because they apply to studies of other traumatic life events.

First of all, as mentioned many times in this book, the majority of studies on traumatic events is conducted with patients or other people requesting for assistance. By definition they suffer from disturbances. Findings of these studies do not necessarily provide evidence on coping in general populations. The issue of representativeness can be met only by selecting a group of people irrespective of whether they consulted a professional health care taker.

A frequent shortcoming of studies on traumatic stress is a high non-response rate. In many studies the final research sample is only a small part of the original group. Non-response rates of 50% to 80% are rather common. Quite obviously the number of people that refuse to participate in research should be as low as possible. It is not known how low response rates influence research findings, although there are indications that non-responders are more adversely affected after a serious life event. Their refusal to participate is likely to be connected with psychological defenses, such as denial and avoidance of the experience and its consequences. In his study of employees exposed to an industrial disaster Weisaeth (1989c) found that those who initially resisted to undergo psychological examination had significantly more serious posttraumatic disturbances at the follow-up examination.

Thirdly, for obvious reasons, an assessment of the individual functioning prior to the event is absent. One possibility of overcoming this objection, at least in research on symptoms, is using a control group of non-victims, but such a condition is usually absent as well. It is uncertain whether victims suffer more than non-victims, for example, from fears and physical complaints. Veronen and Kilpatrick (1983), however, who have utilized control groups, found distinct differences between victims and non-victims.

Time is a highly relevant factor in traumatic stress research. Several methodological problems have to do with the neglect of this factor of time. The length of the time period between the event and the measurement varies from study to study, and is often not specified. Coping with traumatic events involves a dynamic process developing in time. It is crucial therefore that the time factor be explicitly controlled for in research. It makes a difference whether one analyzes the effects in the first week after a violent crime, or six months later. Knowing this we may be surprised by the vicissitudes of this variable in the selected studies. Although in many studies all individuals were measured after a fixed time-interval, it is sometimes not mentioned at which point in time the interviews took place (Sutherland & Scherl, 1970), or the information is not presented separately for the various points in time when the victims were interviewed more than once (Burgess & Holmstrom, 1974).

There is also the related issue of the retrospective character of the measurements. For instance, some 8 to 41 months after the event had taken place, the former hostages of the Dutch hijacking (Bastiaans et al., 1979) were asked to recall what their reactions had been during the first four weeks following the actions. Not only is memory not perfect but events experienced since may color the image of the first days after the hijacking.

The last time related issue concerns the period to which the measurement refers. The studies that were analyzed in this chapter are rather equivocal in this respect. In one research project the subject was required to indicate which symptoms he or she experienced at the moment; in other studies he or she listed the symptoms experienced during the last week, the last month, the last year, or even the whole period of time after the event. This variation in the target period complicates the comparison of findings.

A longitudinal research plan is optimal, especially when using a control group, because it may then be established when the victims have returned to their normal level of functioning (Veronen & Kilpatrick, 1983). One potential objection is a 'Hawthorne effect': the researcher's intensive attention might stimulate the coping process and thereby reduce the complaints.

The next objection concerns the way of measuring consequences of violence. In some studies subjects were asked to rate the symptoms which bothered them most (e.g., Bastiaans et al., 1979). Other studies employed rating scales that have been checked for validity and reliability. Finally, clinical interviews or assessments by experts were used to determine the presence of the disorder. Several studies mention extremely little quantitative data on psychological reactions after the event, so that magnitude, intensity and the course of symptoms over time remain unclear. The diversity in this area is large (Peterson et al., 1991), which again does not enhance comparability. Some publications do not even give adequate information on the operationalization. Future studies should strive for standardization of instruments.

A subsequent shortcoming is the neglect of individual differences. Intensity and duration of coping may vary considerably. Various social and psychological factors, e.g. age, education, personality traits and personal contacts, play a role in determining the whole of reactions to a shocking experience. Very little attention is paid to such determinants, especially in the area of violent crimes. We will return to these factors in detail in chapter 9.

5.6 A comparison with other traumatic events

To what extent are effects of violent situations and their related coping processes similar to those of other events?

As no explicit research has been conducted on this subject, no empirical material is available. Nevertheless, we may make comparisons, partially based upon several examinations of the similarities and differences (Horowitz, 1976; Luchterhand, 1971), particularly those concerning natural disasters (Frederick, 1980).

The similarities between the symptoms of a violent crime and other shocking events are great: fears, insomnia, irritations, dejection, etc. There is a temporal development with components such as disbelief, re-experiencing and avoidance. Differences also exist; certain symptoms emerge more explicitly in one event than another.

Self-blame

The victim of violence often feels guilty because he was not able to prevent the event. 'Someone else has done something to me, so I must have deserved it'. The victim feels responsible. These feelings are less prominent in natural disasters, in which the people who are struck realize they have had no influence whatsoever upon what has happened. However, according to Frederick (1980), hostility and resentment occur much more frequently after disaster, especially in the form of irritation towards the authorities and of deviant behavior such as looting. The latter statement is questionable in view of the few reports of deviant behavior following disasters and in view of the fact that victims of violence may also be irritated by the attitude of the authorities (Jaspers, 1980).

Although guilt is a reaction that emerges after every shocking experience, it appears to be stronger following violent situations than following a disaster. In this connection, an explanation from attribution theory is important. This theory assumes that people attempt to attribute the occurrence of events to certain causes (Jaspars, Hewstone & Fincham, 1983). It may be regarded as an attempt to get a grip on one's own existence. There is a need for control. The confrontation with an uncontrollable situation is unbearable. Therefore the individual would rather attribute the circumstances to himself than to coincidence (Wortman, 1983).

Although it is true that feelings of guilt emerge more frequently following extreme experiences than appears justified by the situations, according to attribution theory one would not expect to find differences between violent crimes and natural disasters; perhaps one should even expect to find the opposite of the above as disasters are even more uncontrollable than crimes of violence. A more appropriate explanation is that guilt does not stem from the realization of total powerlessness, but rather emerges from the assumption that there was some room for personal initiative in the situation, despite its extreme nature, of which no advantage had been taken: 'if I had paid more attention, it would not have happened'. The victim of violence experiences more control, no matter how little, than the victim of a disaster. This creates an awareness of responsibility and, consequently, the potential occurrence of guilt feelings.

The effects of shocking experiences be better understood if they are linked to the nature of the circumstances (chapter 9). Although disasters and violent crimes are similar in many respects, the difference lies in the presence of an aggressor. We have already pointed this out in the introduction to this chapter. One individual grossly violates another individual's freedom. Violence is an interaction – no matter how inappropriate this term may sound – between the person causing the situation and the victim. In the case of violence, the idea that the victim has contributed to the event is present more strongly than in the case of, for instance, a disaster, in which the cause has no 'human' face. Especially when social factors encourage it, the thought soon emerges in both the victim and the immediate environment that the person was partially responsible. A well-known example is rape: 'if she had not wanted it, she would have been able to prevent it'.

The presence of another person creates the idea (or rather the illusion) of the possibility of control, just as it clarifies the strong feeling of humiliation. The situation implies that the victims of violence are reduced to objects. Self-respect is therefore more likely to decrease following a violent crime than after a natural disaster. Among the crimes of violence there will, in addition, be a difference between the experiences in which there is much contact with the aggressor (rape) and those experiences in which this confrontation is less direct (robbery).

The impact of the social environment

Another difference concerns the role of support from the social environment, which is related to the fact that a violent crime usually concerns an individual and a disaster concerns a group of persons. After the latter event, a phase occurs that is referred to as the 'honeymoon phase' or 'post-disaster utopia' (chapter 4). People are happy to have survived the catastrophe, they talk to each other and resolutely start rebuilding their lives. The victim feels appreciated (Luchterhand, 1971). We see much less of these sometimes nearly euphoric phenomena in violent situations, to some extent after hijackings, in which everyone is glad to have survived, but rarely after a rape. Here too, there is the need to be occupied with the event, but the victim of violence, particularly in the case of rape, has fewer opportunities to do so. She feels ashamed and guilty.

Disturbances following violence are sometimes complicated by the contacts of the victim with the police, newspapers, insurance company personnel and medical authorities. The first goal of the police, for instance, is solving the crime, not assisting the victim. Their contacts with the victim are professional; they have to interrogate them and they have to check or prove their statements. There is, therefore, an imbalance in interests of the police and the victim, and that can be especially hard for the latter one who already feels himself humiliated by the act of violence. This negative influence of the social environment is called *secondary victimization* (Steinmetz, 1984). It implies that victims are revictimized by the work of the authorities and their indifferent attitude towards the needs of the victims resulting in more anxiety, despair, irritation and other symptoms of maladjustment. It is often characteristic for the treatment of people following acts of violence. Another source of secondary victimization is the role of insurance companies. The anger of victims of crime and accidents about continued absence of proper insurance settlements may be great.

Identification with the aggressor

There is a striking phenomenon which primarily emerges in the literature on hostage situations, but which was described much earlier by Bettelheim (1943) in relation to the prisoners of Nazi concentration camps: identification with the aggressor. This phenomenon directly results from the presence of the aggressor, which is direct and long-lasting. Extreme isolation and helplessness place the hostage in a position of complete and utter dependence.

The aggressor's omnipotence may result in the hostage identifying with him. This identification, which is mainly unconscious as a psychological mechanism, brings relief from the fear, because the individual more or less shares in the aggressor's (superior) power. Instead of being a passive victim, one becomes a person who actively manifests himself. As formulated by Van Dijk (1981, p. 196): 'It is a reaction in which one inwardly takes leave of the now currently lost and unattainable world, and settles down in the new one, which powerfully manifests itself.' Thus, the victim will then exhibit positive feelings toward the aggressor and sometimes even negative feelings towards outsiders such as the authorities involved in the solving of the hostage situation (Ochberg, 1980). These feelings may remain when he has come through the hijacking safe and sound.

The question is whether this phenomenon occurs frequently. In the later Dutch train hijackings, they were deliberately avoided by both the hijackers and the hostages. In the next chapter we will conclude that identification with the aggressor by concentration camp prisoners did not occur as frequently as was suggested.

A neglected traumatic experience: the traffic accident

Traumatic stress research has mainly dealt with violence and war. Yet there are other events, the effects of which clearly correspond with the reactions to the extreme experiences discussed in this chapter. A traffic accident is an example of such a far-reaching event. It is also characterized by powerlessness, disruption and discomfort, and it is pre-eminently a situation in which anticipation is not possible.

Unfortunately, the lack of attention for the psychosocial effects of traffic accidents is striking. This may be caused by the fact that the victims as such do not turn to mental health care, but to their physician with specific complaints such as headaches or insomnia. Moreover, the neglect might be considered a sign of unwillingness to confront the frightening side of our comfortable and precious system of transport.

The few existing studies (Foeckler, Garrard, Williams, Thomas & Jones, 1978; Hofman, Kleber & Brom, 1990; Noyes, Hoenk, Kuperman & Slymen, 1977; Thorson, 1973; Walker, 1981) on the psychosocial effects of traffic accidents permit the following conclusions. Immediately after the accident victims experience reactions such as disbelief and an almost 'instinctive' response. This is followed by fears, irritations, preoccupation with the accident, dejection and feelings of shame and guilt related to potential responsibility for the accident. If others have been killed in the accident, there is a mourning process involved, which may become serious if the deceased was close to those involved (chapter 7). All of these effects may interfere with the person's daily life to a lesser or greater extent and in some cases may still be present after months or even years (a similar argument may be made concerning the consequences of other accidents; e.g. Braverman, 1980; Malt, 1989; Tollison, Still & Tollison, 1980)

Due to methodological shortcomings, the results of the few studies should be examined with some reserve. Most research is limited to patient samples. However, in a sample of severe road accidents registrated by the police Hofman et al. (1990) found that one month after the accident about half of the victims showed moderate to severe symptoms of intrusion and avoidance. An average of about ten percent of the victims still suffered from posttraumatic stress disorders six months after the accident.

It is not always possible to establish with certainty that the complaints and symptoms of victims actually result from the confrontation with the accident. The effects initially appear to be primarily physical, so that social and psychological reactions, such as fears and depression, remain in the background. The question is whether the effects described here are psychological reactions to the shocking event or rather the result of organic injuries. Traffic accidents are the most important cause of brain damage, resulting in dizziness, headache, nausea and severe physical injuries. Also mentioned, however, are psychological symptoms such as irritability, loss of concentration, partial amnesia and loss of interest in activities (Minderhoud & Van Zomeren, 1984). Do these symptoms originate in traumatic brain damage or in the emotional shock of the accident and the changes it brings about?

It is difficult to draw the line between the two. After all, organically caused problems produce psychological and social effects. The individual worries about his listlessness and slowness, continually relates his complaints to the accident and finally suffers from fears and depressions.

This means that the effects of accidents should be considered from the perspective of multicausality. Organic factors as well as psychogenic and sociogenic factors play a role (Minderhoud & Van Zomeren, 1984). An important implication is that the diagnostics of the disturbances will have to take place much more explicitly in future research. Many studies do not indicate clearly whether or not brain damage had occurred in the persons studied.

5.7 Conclusions

A violent crime is a far-reaching event that may result in various effects of a primarily negative nature. A development may be discerned in those reactions. The victim initially often reacts in disbelief and bewilderment; he subsequently reorganizes his existence but suffers from recollections, avoidances and emotions for quite a while. Recovery often takes longer than the outside world believes it does. Eventually the victim recovers and the event has been shifted to the background in his thoughts and acts, although it has not disappeared.

Striking effects of violent situations are anxieties (which may remain for a long time and are, among other things, expressed in a feeling of insecurity), disbelief (especially immediately afterwards), feelings of guilt and self-blame, anger, re-experiencing and preoccupation with the event, and reactions of an avoidance-oriented nature, such as moving house. This diversity of reactions clearly indicates that both the behavior and moods of the victim shift considerably.

The phases distinguished in the complex of behaviors and experiences in response to a violent crime must be regarded as an abstraction of a process that takes place over time. The sequential divisions of the effects are integrated in a model which, to a great extent, overlaps the development of the stress response syndrome formulated in chapter 2.

The process mentioned above occurs after every crime. Partly due to many methodological problems, this proposition has not been sufficiently substantiated by solid empirical material. However, the severity and duration of the process and its manifestations depend upon the context of the events, and various social and individual determinants play an important role as well. All of these factors have been insufficiently examined in the studies conducted thus far.

We have discussed several differences between serious life events. It was shown that feelings of guilt, humiliation, social isolation, doubt about one's own responsibility, and identification with the offender occurred more frequently after acts of violence than after other events. These relative differences could be explained in terms of the specific nature of the violent situation, namely the presence of the aggressor.

CHAPTER 6

CONCENTRATION CAMPS

Although the emphasis in this book is primarily on serious life events that occur suddenly and are of a short duration, we examine in this chapter the consequences of extremely long-lasting traumatic situations, in particular World War II concentration camps. Long-term effects of traumatic stress have been studied extensively in the survivors. The literature on this subject has influenced publications on other serious events and situations. The study of human behavior during and after the most extreme events may promote a better understanding of the effects of other shocking circumstances.

This chapter focuses on the psychological and social effects of imprisonment in the concentration camps during World War II. Other extreme, long-lasting situations of violence and humiliation – brainwashing, torture, genocide – are not discussed here. Although these have hardly been subject of psychological research, the impression exists that their effects are similar to those of concentration camps (Lifton, 1979).

6.1 The extreme conditions of the concentration camp

The long-term effects of living in a concentration camp can only be understood when we examine the events during and after captivity. It is therefore necessary to direct attention to the psychological aspects of these experiences. The following outline focuses especially on life in the German concentration camps, which were set up for the extermination of the Jews (Bettelheim, 1943; Cohen, 1953; Dimsdale, 1974).

Life in the Nazi camps was one long series of terrors. From the moment of arrival, the prisoner was abused, humiliated and confronted with his inferiority by means of crude violence. He was reduced to a number. The death threat was continuous. Relatives and friends were separated and led to the gas chambers. Along with many other deported persons, one lived packed together in cramped spaces, with no privacy. Hygienic conditions were poor and winters were bitterly cold. Above all, there was hunger. The prisoners were hardly fed, and they had to perform heavy physical labor. Finally, there was absolute hopelessness.

The behavior of the person who was able to survive beyond the arrival was largely determined by self-preservation. Extreme circumstances and terror required this adjustment. It was hardly possible to maintain prewar principles, norms and rules of conduct. Disruption (chapter 1) occurred to an extreme degree: contact with the old existence had been completely cut off. All behavior was accepted if it maintained self-preservation. Hunger dominated all other needs. Food became the central theme in the prisoners' thought, according to Cohen (1953). The phenomenon described here is often indicated as regression: a return to an earlier stage of development, with increasing primitive emotions, support seeking, helplessness and with a completely changed view of the world (Hugenholtz, 1984).

The pursuit of self-preservation resulted in excessive alertness. Time and time again the prisoner had to attempt to avoid death. By behaving as inconspicuously as possible, he could prevent being selected for the cruelties of the guards; he could become the target of abuse at any moment. This resulted in his 'swallowing,' as is were, the anger and aggression.

Regression was also expressed in a lack of compassion for others. Forced by abominable circumstances, but also as a result of a hardening process, many prisoners were selfish in their behavior (Cohen, 1953). Degraded to a number, many became indifferent to the suffering of others. This is indicated by terms such as 'robotization' and 'psychic closing off'. This emotional numbing resembles a long-lasting form of the phenomenon of disbelief or depersonalization during or immediately after a shock, as mentioned in previous chapters.

The SS and the prison guards were omnipresent in the camp, where everything was arranged in an extremely authoritarian manner. This resulted in an almost childlike dependence of the prisoners, and in some cases in identification with the aggressor. Continually confronted with the omnipotence of the nazis, the prisoner, in his fear and hunger, identified with his oppressor. It must be pointed out that this identification did not include the moral standards of the camp personnel, but solely their behavior and, in particular, their aggressiveness. It is better to speak of conforming; complete identification (also called the master-slave syndrome) occurred rarely (Luchterhand, 1980).

For those who managed to survive, the symptoms described above partially disappeared. Particularly the long-term prisoners accepted more successfully reality and were remarkably more active in their resistance against the camp leadership (Cohen, 1953; Luchterhand, 1971, 1980). Many did not attain this point. Especially during the first few weeks of the internment, people suffered from apathy and sadness at the loss of contacts with relatives and friends. This could result in a phenomenon, repeatedly mentioned in the literature: the Muselman behavior. Numbed by continuous misery, hunger and exhaustion, the prisoner stopped caring about being inconspicuous, reacted like a robot and no longer cared for food. Death was usually the result.

Liberation did not mark an end to the suffering of the prisoners. Many still died after May, 1945 of illness and exhaustion (Eitinger, 1980). The return to society in itself proved to be another shocking experience. Relatives had been murdered, possessions were missing and houses had been devastated or occupied by others. The survivors were treated with indifference. Most inhabitants of European countries suffered under the German occupation. Material losses were tremendous and society had to be rebuild. As a result the recognition of the hardships and losses of the surviving Jews were rather minimal. The survivors often started to realize what they had lost only after the war had ended (Eitinger, 1964). In the camp, driven by self-preservation, they had existed in a state of regression and apathy.

6.2 Sequelae of the concentration camp

For a considerable length of time, academic researchers struggled to find an adequate description of the symptoms in former prisoners. Reasons for this struggle were conceptual bottlenecks, a social climate of denial of the misery of war, and, above all, the abnormal conditions under which the camp survivors had lived. Tas (1946), De Wind (1949) and Friedman (1949) were among the first clinicians who reported the severe aftereffects of the internment in the camps.

Scientific research was hesitantly initiated. Motivated by special legislation in some countries, researchers attempted to find a somatic basis to the problems. Early studies were conducted by Danish physicians (Helweg-

Larsen, Hoffmeyer, Kieler et al., 1952; Herman & Thygesen, 1954). They concluded that prisoners suffered from emotional disturbances, intellectual deterioration, and a considerable physical decline, which pointed to an acceleration of the aging process. The theory of a biological basis became well-known; Eitinger (1964) stated that the camp syndrome resulted from organic brain damage, caused by the injuries and the hunger dystrophy in the camps. He discovered that the occurrence of the symptoms was related to head injuries, illnesses during internment in the camp and to dramatic weight loss.

Others (Matussek, 1981; Venzlaff, 1964) disagreed with this neurological explanation, the so-called 'body loss hypothesis'. It was established that those who had not experienced grave malnutrition but had been deported to a camp, or those who had gone into hiding during the war, suffered from the same disorders years later. At present, the argument that traumatic situations of long duration but without the loss of weight may also produce a camp syndrome, is generally accepted. Eitinger (1980), in later years, modified his proposition and recognized that psychological trauma without brain damage could produce a camp syndrome in all its aspects.

Reviewing the literature on the effects of the concentration camps, it becomes clear that the initial efforts to classify the disorders in somewhat rigidly formulated diagnoses were gradually abandoned and that terms such as (post) concentration camp syndrome (Herman & Thygesen, 1954), KZ syndrome (Bastiaans, 1970) as well as survivor syndrome (Niederland, 1961) were preferred. The specific nature of the problem is thereby recognized.

So just what is the concentration camp syndrome? Two types of definitions may be distinguished. The first focuses on a phenomenological description of the individual's experience; the second focuses on the symptoms and complaints of the ex-prisoners. The former is of a theoretical nature, the latter of a more descriptive nature. Since these definitions are complementary, it is useful to highlight both.

Central themes

The following phenomenological description focuses on central themes in the behavior and experience of the survivors. Hugenholz (1984) points out that life after the confrontation with death and destruction has taken on a different meaning for those who survived the camps. A dimension has been added to their existence. What does this dimension consist of? To answer this question we turn to the work of Robert Lifton (1980), who derived a number of themes from his interviews with survivors of the camps and extreme disasters, such as the bomb on Hiroshima. These themes constitute the essence of the survivor syndrome (chapter 4).

The first theme concerns the *death imprint*. The victim has been confronted with death in such an intense way that the images are indelible. Smoking chimneys, separation from friends and relatives, the sight of the corpses of fellow prisoners, are memories that the individual carries with him constantly. Lifton relates these experiences to one's own fear of death. The individual has realized his own mortality intensely.

Survival guilt is the second theme. Why am I still alive while the others have all died? This guilt feeling is a phenomenon that occupies a marked place in the literature on concentration camps. It is rooted in the extreme powerlessness the individual experienced while in the camp. He could not do anything against the horrors. Often he did not feel anything either: no anger, no sadness or compassion.

The survivor feels responsible for the death of others. In suddenly emerging thoughts and in dreams, he feels guilty. Because the responsibility is not placed upon the Nazis but upon himself, one often speaks of paradoxical guilt.

Lifton names psychological *numbness* as the third theme, pointing to the inability to experience emotions. The absence of emotions that characterize the camp period continues to be present. This mental anaesthesia can be regarded a defense against overwhelming thoughts and emotions related to the camp. It is expressed, among others things, in apathy and in withdrawal from human contacts. It is evident that the already known dimensions of coping with traumatic stress – intrusion and denial – return in all themes of the survivor syndrome.

Moreover, this syndrome contains the *complete reversal of norms and values*. This means that the survivor has been confronted with a world completely opposite to his own to such an extent that he can scarcely return to a normal existence. This is the same phenomenon as the acute disruption we discussed in chapter 1, the only difference being that it now occurs to an extreme degree. Lifton primarily regards it a moral issue. Murder and terror were so omnipresent in the camps that they became the norm. The face of morality changed dramatically. But this was only one of the many reversals in camp life. The disruption went further. The prisoner had to leave behind the old, familiar structures and accept a new pattern of norms, values and rules of conduct.

According to the Dutch psychiatrist Bastiaans (1974), a separation process occurred. The prisoner was cut off from his own environment, his rights and possessions by the violence of war, which rendered the restoration of healthy communication or a healthy equilibrium practically impossible afterwards. Bastiaans therefore views the camp syndrome as a permanent, chronic blockage of human relations.

As the next theme, we wish to examine the *disillusionment of returning* after the war. Although Lifton does not mention this theme, its importance is shown in the work of Keilson (1979) who described the sequential traumatization of Jewish war survivors. Postwar society was not like the victims had imagined it would be. In the camp the hope was cherished that the world after the liberation would be a different world, in which oppression and injustice no longer existed. Reality, however, was totally different.

As a final theme, Lifton mentions the *search for the meaning of the events*. This is a more general theme that is essential to all the mentioned aspects of the survivor syndrome. Confronted with omnipresent death and with a complete reversal of values and norms, the survivor searches for the reasons behind what happened to him. He attempts to find meaning in all the horrors, so that life will not be unbearable (Frankl, 1959). We will return to this theme in detail in chapter 8.

Symptoms of the concentration camp syndrome

The themes mentioned above are expressed in a diversity of symptoms. Although some of these have already been mentioned, it is useful to name them separately, because they summarize, in a descriptive way, the problems of the survivors. The following list is a summary of the symptoms of the camp syndrome, presented in various publications (Bastiaans, 1957, 1974; Chodoff, 1966; Eitinger, 1964; Grübrich-Simitis, 1979; Hoppe, 1971; Krystal, 1968; Niederland, 1961). The concentration camp syndrome is a constellation of chronic symptoms, but not all symptoms are at all times present.

Fears, which occur in many forms: startle responses, avoidances (phobias) of situations that arouse recollections of the camp; a vague feeling of approaching doom and nightmares.

Chronic dejection and despair, accompanied by numbing, passivity and lack of interests. The joy of life has disappeared; pleasure no longer exists.

Irritability. As it was absolutely impossible to express anger in the concentration camp, the survivor does not know how to handle his own aggression. It has been bottled up, and every now and then it emerges in the form of rage, irritations and restlessness.

Recurrent intrusive memories. The individual reexperiences the events in the concentration camps time and time again. The survivor is preoccupied with the war.

Reduced psychological resilience (sometimes labelled as asthenia), which is expressed, for example, in the inability to concentrate and memory loss, a lack of vitality and problems with functioning in daily existence.

Nightmares and sleep disturbances. Recurrent distressing dreams that recapitulate real experiences are very common.

Psychosomatic complaints such as dizziness, locomotive disorders, stomach and intestinal aches and increased fatigue. It is often a form of hypochondria. As physical survival was necessary in the camp, the survivor is still very much concerned with his body. Mortality and illness rates, especially due to cardiovascular diseases, are higher for camp survivors (Eitinger, 1964).

These characteristics of the camp syndrome may be profound and chronic. Especially the chronic despair and the recurrent intrusive images of the war may manifest itself in more or less permanent changes in the images that the survivor has of himself, his body and the environment. We mentioned, for example, the inability to look at the future and the continuous awareness of vulnerability. Permanent personality changes may be an essential part of the camp syndrome.

Social effects

Psychological disturbances of concentration camp survivors are strongly emphasized in most publications. Social consequences, however, are not given much attention. These effects are regularly traced back to the psychological reactions, but this is not totally justified. As a result of the long-term stay in the camp, the social life of survivors is often problematic (Matussek, 1975). In turn, the social effects evoke psychological reactions. A feeling of emotional isolation is often present. The former prisoners have trouble entering into relationships.

Many survivors are said to have escaped into work or marriage in the first years after the war, and this led to difficulties in the long run. Several authors (Ostwald & Bittner, 1968; Venzlaff, 1964) point out the hasty marriages that were unhappy and frequently remained childless. The survivor attempted to solve the loss of his relatives and the loneliness by marrying, in several cases another survivor.

It is suggested that survivors are more successful in their jobs than other people are (Bastiaans, 1974; Ostwald & Bittner, 1968) because they cling to their work. Ostwald and Bittner observed an almost compulsive orientation towards success and making money. By working day and night they attempted to regain the security and confidence that had been lost. However, psychological symptoms, such as despair and fear, were not absent in these socially successful persons; the compulsive pursuit of outward signs of success provided little satisfaction.

Whether actual success in work applies to all camp survivors is questionable, particularly in view of the research findings of Eitinger and Strøm (1973), which indicate more work problems in this group. Social adjustment probably proceeded well in the first years, but eventually more disturbances emerged as a result of the one-sided emphasis on job and career, parallel to what happened in married life.

According to Bastiaans (1974), we can distinguish a sequence in the coping processes that develop in the survivors. After the war, adjustment to everyday life took place. During this period, the individual particularly concentrated on social career and family life. There were relatively few complaints. Bastiaans and several other authors (Niederland, 1971) speak of a latency period that was characteristic for many survivors and that could last for years. The term latency phase should, however, be used with caution as it evokes the unwarranted suggestion (chapter 5) that there was a symptom-free period. Instead, the hard work and the hope for a new existence pushed the memories of the camp to the background.

As soon as new disappointments occur, or after the climax of the successful social adaptation has passed, the symptoms increase. On the one hand, chronic over-activity and aggression emerge, while, on the other hand, despair and listlessness occur. This phase is accompanied by psychosomatic, particularly cardiovascular, symptoms. According to Bastiaans (1974), it is characteristic that the emotions surrounding the camp are suppressed, but that this is never completely successful. Suddenly the memories of the camp surface again. Here we once again find the alternation between denial and intrusion that is so characteristic of traumatic experiences.

The final phase

The last phase of the camp syndrome is not very hopeful, according to Bastiaans (1957, 1974). Eventually the capacity for control and repression diminishes, partly due to the early aging that results from the physical hardships of the camp. The adjustment yields to exhaustion, which occurs in the form of a return to the alarm state in which the prisoner found himself in the camp. He is chronically fatigued and is alerted time and time again by circumstances that remind him of the camp, such as news reports on terrorism.

However, the pessimistic view of Bastiaans has proven wrong in many cases. Such an illness model of the consequences of concentration camps may be influenced by the exclusive interest in patients and psychiatric disorders. Many survivors have managed to cope with their experiences; they were in good health and did not break down. In an Israelian study (Antonovsky, Maoz, Dowty & Wijsenbeek, 1971), it was found that 40 percent of a population sample of female camp survivors were rated as being in excellent or quite good health for women of their age.

Recent studies

In the past few years controlled studies concerning adjustment of survivors of the Holocaust have appeared (e.g. Dor-Shav, 1978). Dasberg (1987) reviewed them and concluded that survivors may have no more somatic impairment than age-matched controls. However, they are more vulnerable to the stresses of life, they are more depressed and they are less optimistic. These conclusions are confirmed in a study by Harel, Kahana and Kahana (1988), who add that they found less differences between survivors and controls than would be expected on the basis of the clinical studies of Holocaust survivors. Furthermore, these authors proved with their well-balanced research that factors such as marital status, physical health and income were more important predictors of psychological well-being in the advanced years of life than whether or not people suffered in World War II.

6.3 Transgenerational effects

The children of Jewish parents who survived World War II have become the focus of attention, both in The United States as well as in Israel and in The Netherlands. It is often mentioned that the war experiences of the survivors put a strain on their child rearing capacities. Because of the many losses in war time survivor parents could not adequately comply with the needs and wishes of their child.

In a number of clinical descriptions of children in families of World War II survivors (Freyberg, 1980; Sigal, Silver, Rakoff & Ellin, 1973) it appeared that separation from the parents and the development of individual autonomy were not facilitated. The growth of the child's independence was felt as another loss (Barocas & Barocas, 1980; Freyberg, 1980). It was difficult for the parent who suffered from traumatic war experiences to accept and stimulate the child's ambivalent wishes to become independent, while at the same time needing the encouragement of the parents. Not being able to support a child to become independent might lead to an 'anxious attachment' between parent and child (Bowlby, 1980), manifested in the difficulty with accepting separations, in overprotection by the parents, in frustration and rage, and feelings of guilt and shame (Russell, 1980).

Memories of the war experiences disrupted family dynamics (Freyberg, 1980; Russell, 1980; Sigal et al., 1973). Parents were found to have problems with intimacy in relation with their children. They were often so occupied by the reminiscences of the war that they could not be empathic to the emotional needs of the child. The intrusions of the war experiences made the parents psychologically vulnerable and they turned to their children for support. As a result of this process children of Jewish war survivors felt responsible for their parents and for the parents' well-being. The phenomenon of parentification, a reversal of roles in the family, is often mentioned in the literature on Holocaust survivors (Danieli, 1982). It has also been stressed that survivor parents expect their children to perform maximally so that they did not survive their ordeal for nothing. This might led to an emphasis on high intellectual achievements (Begemann, 1991).

Another aspect which affected the children of survivors is the conspiracy of silence (Danieli, 1982). The World War II experiences of the parents were hardly discussed within the family.

Clinical investigations indicated that problems mentioned above might lead to a reduced psychological well-being and more psychopathology in later life (Russell, 1980). In particular, feelings of depression and guilt, and difficulties in the expression of emotions are found (Nadler, Kav-Venaki & Gleitman, 1985). Excessive commitment, irritation and mistrust are reported as manifestations in relationships with friends or partners (Barocas & Barocas, 1980). The reduced psychological well-being might cause a greater need of professional help.

Empirical studies of nonclinical groups of children of Jewish survivors using standardized questionnaires and/or a control group design, in general, did not confirm these results (Leon, Butcher, Kleinman, Goldberg & Almagor, 1981; Rose & Garske, 1987). However, empirical studies using standardized questionnaires have often been criticized for reasons that these instruments are not sensitive enough for grasping the specific intra-psychic issues of the so-called second generation (De Graaf, 1975). These contradictive findings as well as methodological problems cast serious doubts on the transgenerational effects of war experiences.

In a Dutch study a representative sample of second generation Jews was compared with a representative reference group (Eland, Van der Velden, Kleber & Steinmetz, 1990). The groups were interviewed in a semi-structured way by experienced psychotherapists. The shortcomings of clinical case studies were avoided by adding a reference group into the design, by studying random samples, and by utilizing structured instruments. In this way a merely impressionistic description of problems in a patient group could be avoided.

This quantitative research of differences between children of Jewish war survivors and a reference group confirmed several transgenerational effects mentioned in clinical studies, even in a random sample of subjects. Children of war survivors considered their childhood as characterized by more feelings of guilt, more feelings of shame and more problems with separations in the family. Disrupted family dynamics were manifested in parentification, problems with aggression, and overprotection by the parents. Current social relationships of the Jewish second generation were hardly affected by the war experiences of the parents. Compared to the reference group, children of Jewish survivors had a more difficult and more complex youth that still bothers them but that is not generalized to all aspects of their current life.

It is relevant to expand the range of research on transgenerational effects of war experiences. Nearly all studies have been conducted with Jewish survivors. The large group of the children of parents who lived in the former Dutch Indies and who were kept in Japanese camps during World War II (Doreleijers & Donovan, 1990) is another example of the so-called second generation. Their cultural background is different. An investigation into the similarities and differences between the groups of war survivors and their children would be meaningful.

6.4 Conclusions

The psychological disorders resulting from the cumulation of long-term stress factors in the concentration camp and afterwards are frequently indicated by the term 'concentration camp syndrome.' This is a rather complex syndrome, primarily characterized by a preoccupation with death and violence, survivor guilt, all types of dejection and emotional numbness, difficulty in dealing with aggression and psychosomatic complaints. Because the survivor adjusted to the extreme conditions only after considerable time had passed, the various symptoms from that period – psychological withdrawal, physical orientation, hyperalertness – may still be present. The effects are often extreme in duration, gravity and frequency and may have resulted in permanent personality changes.

Although the concentration camp syndrome is used to designate the psychological problems of concentration camp prisoners, it also applies to similar disorders resulting from other lengthy and extreme situations of terror, humiliation and hopelessness. The term also acquires a broader significance in Lifton's work, which provides it with an existential value so that, in fact, every survivor suffers from the camp syndrome. The concentration camp syndrome sometimes becomes a general, almost metaphorical indication of the problems suffered by survivors of lengthy, extreme situations, in particular because it seems impossible to draw the line between normal suffering and pathological forms of coping. 'Die Ermördung von wievielen seiner Kinder muss ein Mensch symptomfrei ertragen können, um eine normale Konstitution zu haben?' (Eissler, 1963, p. 241).

CHAPTER 7

LOSS

Death is a subject that received very little attention in academic psychology until approximately the early seventies. There are several reasons for this lack of interest. Death is a broad, unfathomable subject that is not easy to study, certainly not in a psychological tradition dominated by laboratory research. Death is a subject that was taboo until recently, and may still be.

In recent years, however, social scientific articles have been regularly published on all types of problems concerning death: euthanasia, dying in general, thinking about death, suicide, terminal care and the death of a loved one. This chapter deals exclusively with the last subject: the death of a loved one and the survivor's subsequent period of mourning.

As with so many psychological topics, Freud was one of the first researchers to address the issue of loss. Together with Breuer (1895), he described the case of Anna O., who experienced serious psychological problems after her father died. In 'Trauer und Melancholie' (1917) he outlined certain characteristics of the mourning process that are still considered essential in the most recent literature. In particular Freud aptly described the continuous concern with memories of the deceased and reexperiencing his death.

Another important contribution was made by Lindemann (1944) who, in a study that has become a classic, described the sorrow of the relatives of the 101 fatal victims of a fire in the crowded Cocoanut Grove nightclub in Boston. In this study, Lindemann attempted to clarify the difference between normal and pathological grief.

The research conducted by British psychiatrists Bowlby and Parkes has been a significant contribution to the field. Parkes, as co-worker and colleague, was strongly inspired by Bowlby's developmental psychological 'attachment theory'. A theory which Bowlby has been formulating since the forties in his studies on attachment, separation and loss in the relationship between parent and child. A great deal of the empirical research conducted by Parkes, in turn influenced Bowlby, as may be seen in the third part of his standard work 'Attachment and Loss' (1980). The theoretical insights provided by these British researchers are a combination of ideas taken from ethology, cognitive psychology, and particularly from psychoanalysis. We will discuss the works of Bowlby and Parkes extensively. These authors have described in detail the development of the grief process after a serious loss, without losing sight of the complexity of the symptoms.

Previous chapters mentioned that much of the research on traumatic events is rather incohesive and lacks a sound theoretical foundation. This cannot be said of the research on bereavement. It is characterized by some fine empirical studies (Parkes, 1971; Glick, Weiss & Parkes, 1974) and by excellent theory building.

Grief reactions do not occur exclusively following a death. Separation from a loved one or departure to another country may evoke similar emotions. However, this chapter will refer to the effects of the death of a loved one, specifically the death of a spouse – a frequently occurring event, because most research has been done on this subject.

After a short definition of terms, this chapter will deal with the various phases in coping with the loss of a loved one. We will discuss pathological or atypical forms of grief. Next we will discuss several social effects, and outline influences relevant to the mourning process. These factors, such as cultural rituals and the background of an ambivalent relationship, will be discussed more extensively in chapter 9.

7.1 Definition of terms

The English language distinguishes between psychological and social reactions to a serious loss, and this distinction is clear in the use of two separate terms. 'Grief' is a biologically founded pattern of physiological and psychological reactions developing along set lines. 'Mourning' refers to the loss-related reactions that have been prescribed by society. Averill (1968) has particularly emphasized this distinction.

Yet such a precise delineation of both terms seems too stringent. Psychological and sociocultural factors cannot be regarded separately. Socially determined rules influence the experience of a great loss, while the observation of cultural mourning rituals will always be accompanied by grief and sadness to some extent. It is for this reason that Parkes and Weiss (1983), for instance, define both terms in a more general way. 'We prefer to reserve the term 'mourning' for the observable expression of grief and to use 'grief' as the term for the overall reaction to loss' (p. 2). Averill (1979a) later also modified his strict delineation of these terms.

In addition to grief or mourning, we will also frequently use the word sadness. Sadness is an emotion, an expression of grief, that emerges due to the loss of a person, an object or other important matters. According to Frijda (1982), loss in itself is not a sufficient condition for the emergence of sadness. The loss must be perceived as final, i.e. irreparable. Frijda refers to this quality as finality. Sadness is thus an emotion that belongs to the symptoms related to grief. The situation of the loss of a loved one as a result of death is indicated by the term 'bereavement'.

7.2 Phases in the coping process

The loss of a loved one is a complex phenomenon. The death of a partner is a loss on many fronts and also a transition to a new situation. Coping with such a loss is gradual and complicated process. It requires time and energy. Researchers have attempted to identify phases in this process. Although there are several classifications of phases (Lindemann, 1944; Ramsay, 1977), the division by Bowlby (1961, 1980) and Parkes (1972) is the most widely known. They distinguish between four phases:

1. bewilderment ('numbness');
2. yearning and searching for the deceased;
3. disorganization and despair;
4. reorganization.

In this chapter we will discuss the reactions to a serious loss based on this classification. Like the phase model presented in chapter 5, these should be regarded as 'prototypical': not everyone experiences the phases to the same extent.

The phase of bewilderment

Bowlby and other authors (Averill, 1968; Lifton, 1979) speak of 'numbing' with regard to the first phase, but the term bewilderment more adequately reflects that this phase concerns not only emotional numbness, but also disbelief and pain. This phase begins when a loved one has died and lasts anywhere from several hours to one week. The surviving relative feels numbed by the loss. He is unable to realize it, is confused and carries out activities almost automatically. It is as if the event were not real, as if the person is dreaming.

Often the survivor impresses the outside world as a strong person. A common example is that of the widow or widower who bears up bravely during the funeral and comforts friends and acquaintances of the deceased.

For some of the bereaved, bewilderment is not only accompanied by this numbness, but also by vehement emotions. One is not capable of accepting the totally new situation, and one is tense. In such cases, expressions of intense grief or sometimes anger are not unusual.

This first phase is very similar to the symptoms immediately following other far-reaching events that have been discussed in previous chapters. Therefore, we will not address this subject any further at this point.

The phase of intense yearning and protest

This phase, which may last several months to one year, is marked by the strong desire of the grieving person to search for the other, to protest against the loss and to harbor the short-lived illusion that he has found the other person.

This phase in the grief process starts several hours or days after the other person has passed away. The individual begins to register the loss and experiences surges of intense grief ('pangs of emotion'), which are accompanied by crying, insomnia and all types of stress reactions.

The individual yearns intensely for the lost one and attempts to find him (Parkes, 1972). Although he is knows rationally that the deceased will not return, he experiences a strong drive to search for him.

This preoccupation with the deceased indicates an attempt to re-establish the bond with the dead person, according to Parkes and Bowlby. The individual desperately searches for reunion with the lost loved one. In this part of the theory, Bowlby has been particularly inspired by Darwin's description of this searching behavior in animals (also Averill, 1979a). Bowlby's work extensively describes the reactions of small children to a separation from their parents. He observes strong similarities between a surviving partner searching for the deceased and a child's attempts to find his mother again (and his protest against the separation). From a rational point of view, this search is a hopeless matter, as the deceased will not return.

Nevertheless, the sadness is often lessened by a sudden sensation of having found the other.

Rees (1971) interviewed approximately 300 widows and widowers from Wales; 39% sensed that their deceased partner was constantly present and 14% experienced hallucinations of his or her presence. This 'finding' of the deceased occurs in dreams as well. The individual reports clearly seeing the deceased. Searching and finding are therefore closely related. 'Thus a widow may be preoccupied with a clear visual memory of her husband; at one moment she is anxiously pining for him, and a moment later she experiences a comforting sense of his presence nearby; then something reminds her that this sense is only an illusion and she is pining again' (Parkes, 1972, p. 60).

In the studies conducted by Rees and Parkes, the widows who sensed the presence of their partners, felt lonelier and missed their partners more than those who did not report such illusions of 'finding'. On the other hand, however, they felt supported by the 'presence' of the deceased and did not suffer as much from sleeping disorders. 'Finding' is therefore regarded by Parkes as a form of easing the pain of the death. Other forms of mitigation are:

- Not believing the other has died. However, prolonged denial is rare.
- Avoiding thoughts of the loved one; avoiding situations and people that bring him to mind. Two thirds of the London widows interviewed by Parkes admitted to this avoidance-oriented behavior, especially during the first month following the loss.
- Selective forgetting. It takes time to put the memories of the other in order. The harder one tries to evoke the image of the other directly after the loss, the more difficult it becomes. Some time must pass before the countless fragmented memories of the other can be integrated into a whole.

It should be noted that all three forms of mitigation are connected to warding off the awareness of the loss, whereas in the beginning of this section the emphasis was on preoccupation with the death. These opposite symptoms alternate.

Many authors emphasize the almost continual alternation between avoidance and preoccupation. In chapter 2 this is described as a central element in the reactions to serious life events. Freud also described the painful process of searching for the lost loved one, being reminded constantly of his death, and attempting to avoid this awareness.

This alternation is regarded as useful. The defense has a useful function, as long as it is not too strong. It provides the opportunity to gradually integrate/incorporate the awareness of the new and unpleasant situation (Parkes, 1972).

Anger is also characteristic of this period in the grief process. It may manifest itself in irritability, bitterness, anger toward doctors and relatives, and sometimes toward the deceased. The surviving partner feels abandoned and lost. Tantrums are not unusual and are sometimes directed at well-meaning persons who want to offer comfort. Feelings of anger are strongest during the first month after the death (Parkes, 1972). Based on his theoretical approach, Bowlby (1980) finds this understandable. The survivor seeks to be reunited with the deceased. He does not want well-meant confirmation of the loss from an outsider, but rather assistance in finding the other person. Later in this chapter we will see that failure of this anger to appear is an indication of complications in the coping process.

To summarize, the following symptoms are the most important characteristics of the second phase in the mourning process:
- restlessness;
- preoccupation with thoughts of the deceased;
- attention for the aspects of the situation that are related to the deceased and/or which suggest that the other person is still alive;
- actual searching for the lost person in the environment;
- crying; intense yearning for the lost person;
- anger, irritation, accusations and ingratitude toward others.

All of these symptoms express the urge to re-establish the bond with the deceased. One should not forget, however, that such symptoms are often short-lived. The anger emerges in spurts as memories surge up suddenly; all at once the widow believes she hears her husband's voice. These periods are called 'pangs of grief', which start several days after the loss and reach their climax 5 to 14 days afterwards. They disappear very gradually (Parkes, 1972). During this phase of mourning other symptoms include:
- intense grief that a loved person has died and will not return;
- attempts to avoid memories of this death and the subsequent grief.

The phase of disorganization and despair

Some time after the loss, the feelings of despair and loneliness intensify. The surviving partner's searching and yearning for the lost one diminishes, as does his avoidance of accepting the finality of the loss. The individual realizes that the other is gone from his life for good. Now the fact that all kinds of activities must take place without the other begins to sink in. This realization often develops in spurts and starts. For example, a widow wants to do something and she automatically assumes that her husband will be involved; only then does she realize he is no longer there and that her actions are in fact inadequate (Bowlby, 1980). This results in feelings of dejection and despair. The old patterns of thinking, feeling and acting must be abandoned, before there are any new patterns. Thus the name disorganization is given to this phase in the grief process. Depression prevails, accompanied by apathy and loneliness. The widow withdraws from social activities to a greater or lesser extent. She avoids those who make demands upon her and relies heavily upon relatives and close friends. She undertakes little (Bowlby, 1980; Parkes, 1972).

There is no sharp boundary between this phase and the preceding one. The transition is very gradual. Moments of intense yearning and despair often alternate: 'There is no sharp end-point to yearning, and pangs of grief can be re-evoked even years after a bereavement' (Parkes, 1972, p. 87). In time, however, the yearning and anger become less intense, while dejection becomes more prominent. This gradual transition also applies to the distinction between the third and fourth phases in the grief process.

The phase of reorganization

While she feels despair, the widow already realizes that certain ideas, expectations and activities are no longer relevant. She gradually becomes used to the facts; she is a widow and no longer a wife. Her loved one will not return. Gradually she becomes more active socially. She attempts to deal with the new situation. Different patterns emerge in her thoughts, feelings, and actions, yet these are accompanied by recurring feelings of loneliness and sadness.

This reorganization has been described by several authors in terms of a theoretical framework in which cognitive processes are central (Bowlby, 1980; Parkes, 1972; Van Uden, 1988). The individual must redefine both self and situation, which is primarily a cognitive process. The survivor is aware of the altered circumstances; the inner representation of the self and the world are adapted to the changes in the circumstances. This concept is discussed in chapter 8.

This last phase in the grief process, which begins approximately one year after the loss, can be clarified with an approach inspired by role theory (Brom & Kleber, 1988). The individual begins to take on new roles. In marriage, the husband and wife fulfilled close, interdependent roles. The husband, for instance, may have simultaneously functioned as a lover, father, husband and administrator, and the wife may have had related roles.

These roles were probably complementary to a great extent (for example: one spouse took care of the household, the other earned income for the family). Many expectations and rules that helped to regulate daily life and solve practical problems were defined by these roles. Therefore, a tight network of expectations, beliefs and behaviors existed.

The death of one of the two partners drastically ends this network. This is very concretely expressed when certain daily routines suddenly become meaningless. The individual no longer has to set the table for two. Role expectations and related behavior must be changed. A widow or widower must arrange a way in which to earn income, to manage the household and to raise the children. It is important to point out these social aspects, as mourning is a process in which all kinds of social changes take place. In line with Parkes (1972), we can distinguish four possible effects of the loss of a partner:

1. the roles of the deceased remain unfulfilled;
2. a replacement for the deceased is found;
3. the roles are taken over by the person in mourning;
4. the family, insofar as it exists, falls apart.

With the loss of a loved one, not only does the widow or widower lose the fulfillment of that person's roles, but he or she is also confronted with a new position in society. Others approach the mourning person with different views and expectations. An individual is relabeled as single. He must build a new identity and at the same time abandon the old one.

It is striking that widows and widowers will strongly sense the presence of the deceased long after they have recovered from the loss. In several studies results showed that they frequently thought of the deceased, sometimes even years later (66% of those interviewed in the study by Glick, Weiss and Parkes, 1974), and that they sometimes forgot that their loved one had died (25%). Sometimes they talked to the lost person (Rees, 1971). The majority of the people interviewed reported that these experiences were very comforting.

These studies also reported that the deceased is present in the dreams of approximately half of the widows and widowers. These dreams are often very realistic, except that the deceased is often seen as young and healthy (Gorer, 1965). However, there is often something in the dream to indicate that all is not well. Sometimes, the final days of the deceased are replayed in dreams.

Bowlby (1980) believes that the deceased partner is experienced as a comrade who accompanies the survivor everywhere, or who exists in some familiar place. In Bowlby's view, this indicates that attachment remains. In this respect, Bowlby is diametrically opposite to Freud (1917/1985), who interpreted the mourning process in terms of the withdrawal of the libido of the loved person. For Freud mourning implied the severing of bonds with the other.

In isolated cases, the awareness of the presence of the other may lead to the realization of resembling the deceased more closely, or even to the feeling that the other is inside the bereft person. The idea of the partner being inside one is a passing phenomenon. Other examples of identification are also usually not complete or permanent.

An alternative to phases

Phases are illusions, but useful illusions. Their purpose is to structure the chaotic diversity of the various reactions in time. Phases are heuristic tools.

However, no individual widow of widower will go through the successive phases exactly as the model predicts. Some people skip certain phases, other people remain in a phase for a very long time, and again other people oscillate back and forth between phases. Moreover, probably most individual coping reactions are hard to fit into the model at all. Empirical studies have confirmed several aspects of the separate phases we have described, but no study has yet proved the whole model to exist (Wortman & Silver, 1989).

One has to realize the risks of these models in practical mental health care. Models of phases have become not just a description of reactions to bereavement but a prescription for adequate coping with loss. They are used as a measuring stick along which the success of the process of coping can be evaluated. The result is that individual differences in the duration, intensity and nature of grief reactions are neglected.

Some authors (Ramsay, 1977) prefer the concept of components instead of phases. They point to the same reactions as Bowlby, such as despair and anger, but instead consider them as elements that one has to go through. They do not postulate a sequence in time.

Another alternative is the approach of Worden (1982). Mourning is seen as involving basic tasks that require effort. The grieving person must accomplish these tasks in order to complete mourning. Again, we encounter the notion that grief is necessary and that a person must accomplish grief work. These tasks do not follow necessarily a certain order, although there is a suggestion of some succession in the definitions. The four tasks of mourning are:

1. to accept the reality of loss;
2. to experience the pain of grief;
3. to adjust to an environment in which the deceased is missing;
4. to withdraw emotional energy from the deceased person and reinvest in another relationship.

Grief as a normal process

In the view described here, grief is a 'normal' psychological process that takes place over a certain period. It was already pointed out by Freud (1917) that grief is not a pathological condition but an adaptive process which requires no intervention. The individual detaches himself from a loved one and this 'grief work' – a term first employed by Freud – is useful. Averill (1979a) and Bowlby (1961) even regard the grief process as biologically functional. Some characteristics (e.g. searching) are relevant when a loss is not irreparable; others (e.g. grief and defense) are functional because they evoke comforting behavior in the group members, or because they protect the individual from too much grief.

The duration of a normal bereavement process cannot be sharply defined. It is true that the most intense reactions will have disappeared after about six to twelve months, but many effects remain present in a less severe form for *at least* one or two years after the loss. They may also suddenly return years later. Death is an event that puts a severe physical and psychological strain on a person, so that the coping process can take quite some time.

Grief is a normal and healthy reaction to an overwhelming loss. Moreover, it is not only normal, but also a necessary phenomenon. The individual must in some way cope with the death of a significant other. The avoidance of grief results in psychological problems (Worden, 1982). This leads to the subject of the following section: atypical forms of mourning.

7.3 Pathological forms of grief

Not everyone manages to cope with the death of a loved one. It is not uncommon for people to suffer for years from anxieties, self-reproach and preoccupation with thoughts of the deceased. In addition, the reactions to the demise can be excessive, as in the case of a prolonged depression. These reactions can be called pathological or atypical grief. Which forms of disturbed coping processes mark the difference between 'normal' and 'pathological' grief?

Various classifications of atypical grief have been established. Two of them will be presented here. The first is based on a review of the psychological and psychiatric literature on bereavement problems; the data are primarily based upon experiences with patients. The second classification is derived from a longitudinal study on a random sample of mourning individuals.

Bowlby (1980) concluded that there are two forms of pathological mourning. Both can reach various degrees of severity; a less severe case is certainly difficult to distinguish from normal mourning.

The first category is chronic mourning, which is characterized by intense and prolonged emotional reactions to a loss, a great deal of self-blame and anger (anger is often directed both at the self and at others simultaneously), depression, anxieties and little sadness. It is as though the mourner is stuck in the second phase of the grief process: a desperate search and yearning for the lost person. The persistence and intensity of the symptoms over a long period of time differentiates chronic grief from normal grief.

Bowlby defines the second type of pathological mourning as prolonged absence of conscious mourning. This delayed form of mourning has been described extensively in clinical literature. It seems as if the individual has coped well with the event, but then he suddenly becomes depressed. The person often suffers from vague physical and psychological symptoms that he does not relate to the loss. It is as though he has remained in the first phase of the bereavement process; emotional numbness. As mentioned earlier, this phase does not normally last for more than a few days. With prolonged absence of conscious mourning, however, this phase lasts much longer, sometimes for many years. The symptoms (tension, headaches, insomnia, etc.) indicate that the process of coping with the loss has not been completed. The cheerfulness of these patients often comes across as exaggerated to the outsider.

It is difficult to detect this variant of pathological grief. It is rather a contradiction to state that the absence of grieving is a form of pathology. Is the bereaved person really denying loss or is the loss not significant for him? Wortman and Silver (1989) concluded that there is little empirical support for the view that those who fail to exhibit early distress have latent pathology and will show significant difficulties. Several studies report a strong relation between distress-scores shortly after the loss and the level of follow-up scores.

This is a matter of definition. Wortman and Silver identified absent grief as an absence of distress, while Bowlby wrote about an absence of conscious grieving. There are subtle clues that distinguish between the pathological condition of grief and the condition of no grief at all. Expressions of sympathy are not accepted. Episodes of irritability or tension occur from time to time (Burnell & Burnell, 1989). Bowlby mentions a striking aspect of these individuals; they often compulsively attempt to take care of others. Even in a situation of a widow or widower with apparently good adjustment there may be signs of distress and somatic complaints that are difficult for an untrained observer to detect. Yet we agree that research data suggest that this variant of pathological grief, even in this more sophisticated definition is not very common (Wortman & Silver, 1989).

The second classification we present here was developed by Parkes and Weiss (1983) on the basis of a secondary analysis of the data of the Harvard Bereavement Study, which included thorough interviews with approximately 70 widowers and widows. It is important that these researchers not only designed a classification, but clarified the causes of the bereavement problems. In addition, this study has a decidedly longitudinal character. Using various interview questions and standardized questionnaires, the researchers interviewed participants a total of four times – after two weeks, eight weeks, 13 months and two to four years after the loss. Probably no other research has been done in the area of traumatic events in which measurements took place at so many various points in time and over such a long period of time.

The dichotomy in Bowlby's work as well as earlier publications by Parkes (1972), Maddison and Walker (1967) and others, has been replaced by Parkes and Weiss with a trichotomy of pathological grief: unexpected grief, conflicted grief (comparable to the delayed grief mentioned earlier) and chronic grief.

Unanticipated grief

Parkes and Weiss came to the conclusion that there is an unanticipated loss syndrome, which is characterized by great difficulty in acknowledging the death, avoidance of the confrontation with the loss, self-reproach, despair, withdrawal, preoccupation with thoughts of the deceased, anxiety, and depression. The authors drew these conclusions by comparing those who had lost their partner without any warning with those who had expected the death.

Confronted with a sudden death, those who had lost their partner unexpectedly were unable to comprehend what had happened. They seemed to avoid the unbearable pain. Only after a certain lapse of time did the reality hit these people, followed by intense, deep grief. With an expected death, on the other hand, surviving spouses experienced a direct increase in anxiety and tension after the death, but these reactions were not very intense.

After three weeks, those who had unexpectedly lost their partner primarily experienced disbelief. There was a schism between the cognitive awareness of what had happened and emotional understanding. Those who did not expect their spouses' death reported symptoms significantly more frequently during the first weeks than those who did. These symptoms included anxiety, self-reproach, feelings of guilt and feelings of abandonment. Visits to the grave and social contacts with friends and relatives lessened.

Most of the differences between groups still existed after two to four years and had become even more pronounced than directly following the loss. For those who did not expect their partner's death, symptoms of anxiety, tension, depression and social isolation continued as did difficulty coping with the reality of the situation.

Conflicted grief

Difficulties in recovering from a loss are more likely to occur after a bad marriage than after a good marriage, although one might expect the opposite to be true. Parkes and Weiss asked detailed interview questions about arguments and disagreements during the marriage. On the basis of the responses to these questions, they divided the sample into two groups: many marital conflicts or few conflicts.

After six weeks, those who had rated their marriages as turbulent were less emotional than those in the few conflicts group. They had more social contacts and once again began to behave like available, single individuals. However, after 13 months, this picture changed dramatically. The relationship between the surviving spouse's coping and the level of conflict in the marital relationship was especially apparent after an expected loss. Virtually without exception, those who had a good marriage and who had been expecting the death functioned well. Among those who had lost their partner after a discordant marriage and had prepared themselves for the death, only one half functioned well.

After two to four years, those with few conflicts in their marriage returned to their level of functioning before the death. However, those who had experienced many conflicts in their marriages displayed more depression, anxiety and guilt. They considered their own role in the marriage as negative. Their health was worse, as was their social functioning. Most remarkable was that the yearning for the deceased was stronger than before: 63% yearned for the partner with whom they had experienced many conflicts, years after the partner's death.

Chronic grief

A number of the widows and widowers studied by Parkes and Weiss struggled from the start with a lack of confidence and with feelings of helplessness and indecision. Characteristically, there was an intense yearning for the partner. Parkes and Weiss therefore compared this group with the group that yearned less for the deceased partner. The degree of yearning at the time of the first interview proved to be significantly related to an increase of problems later in the bereavement process. Thus, pathological mourning definitely does not imply delayed mourning.

This incessant yearning was especially present among those who had been strongly dependent upon the deceased spouse. The strong bond between husband and wife was characterized by uncertainty for the person, who had clung to their partner while he or she were alive. When the partner died, the survivor's sense of security and safety disappeared in one fell swoop. The mourner did not know what to do or how to arrange his own existence; his sense of direction in life had disappeared.

After one year, these 'high yearners' became more preoccupied with the loss and more lonely. After two to four years they suffered from markedly more depressions than did the group of 'low yearners'. One third of them still mourned during this stage. Parkes and Weiss view chronic grief as an inability to continue on alone and a 'clinging' to the deceased, upon whom the mourner had been very dependent. Dependence and intense yearning form the essence of this type of pathological grief.

Parkes and Weiss' study offers particularly good insight into the background of problems in the bereavement process. However, the research is characterized by methodological weaknesses. The relationship between the three different types of pathological grief is not clear. When the figures presented by the authors are critically examined, one gets the impression that there is an extensive overlap, which creates uncertainty about the separate character of each syndrome.

The difference between pathological and normal grief is gradual. A few authors (Raapis Dingman, 1975) even refer to the division as arbitrary. Are there a few criteria that could define this boundary more clearly?

According to Schneider (1980), time is the most important criterion. When very intense grief reactions last longer than six months, then grief begins to take on a pathological form. The six month period is often mentioned in the literature (Volkan, 1979). It does not mean however, that preoccupation, grief and loneliness have disappeared among normal mourners after six months. Now and then these feelings return, especially at key times such as during holidays, but the person begins to rebuild his life most of the time.

Avoiding reality is one characteristic of pathological grief. The person reasons as though the lost one is still present. Examples of not acknowledging reality include continuing to speak of the deceased in the present tense or being hostile toward someone who refers to the death.

Another characteristic is that the pathological mourner remains preoccupied with the deceased. The dead partner is still taken into account: her clothes are kept clean, the table is still set for him, etc.

Finally, in pathological mourning, the mourner avoids situations that are reminiscent of the loss. The widow or widower avoids contact with others as well as discussing the death of the loved one. This behavior is justified by remarks that this does not serve any purpose or that he or she does not want to evoke any pity.

These criteria all suggest that the surviving spouse has remained caught in the first three phases of the bereavement process, as described by Bowlby and Parkes. The often intense emotions, especially from the second phase of grief, have continued to exist so that the reorganization process cannot begin – hence the importance of the time criterion (Volkan, 1979). The loss is still seen as reparable; in particular the separation of cognition and emotion shows that the widow or widower has not yet really integrated the death and its implications.

Depression is often associated with grief. However, it is not correct to use this term in normal grief, because although a mourner experiences distinct moments of sadness, despair and disinterest, he is not continually immersed in a chronic state of depression. A much wider diversity of reactions can be seen in normal mourning than in a depression. Sadness alternates with anger, social activities and more agreeable moments. In mourning, the emotions previously referred to as 'pangs of emotion,' come and go much more frequently than in depression.

Schneider (1980) also clarified the distinction between normal grief and depression. One of the most significant differences he mentions is that a depressed person sees his loss (depression also often begins with a not clearly recognizable loss) as an indication of his worthlessness. The individual's negative self-image is validated. A mourning person more frequently asks himself how the loss could have taken place. He searches for a cause and sometimes arrives at self-blame ('I didn't take care of her well enough'), but changes that opinion of himself under the influence of others. Depression has, however, a life of its own. Depressive symptoms occur regardless of day to day events or triggering emotions (Burnell & Burnell, 1989). The mourner is preoccupied with the deceased and the loss; the depressed person is preoccupied with himself. 'In mourning it is the world that has become pathetic and empty; in melancholy it is the self' (Freud, 1917, p. 76). Depression is, in fact, an inability to mourn.

Depression is certainly not equal to 'normal' grief, however there is a clear resemblance between depression and pathological grief, in particular chronic grief. This is especially the case when the loss is purely or merely a stimulus through which already existing disorders and suppressed hostility emerge (Janis, 1971). Prolonged problems with grief can overlap with a clinical depression, as feelings about one's own helplessness are predominant. There are also implicit differences, as with pathological mourning, with the deceased having a much more central position.

Note that an 'excessive' reaction to a death does not always take on the form of syndromes mentioned above, but that it may also be expressed by various medical complaints and certain behaviors (for example 'acting out,' such as increased sexuality). Such reactions often take place simultaneously.

It is not easy to differentiate between normal and pathological mourning. The boundary between healthy and unhealthy is diffuse. It is not surprising, therefore, that it is difficult to indicate the incidence of pathological grief. Bowlby (1980) notes that various studies employing representative random samples estimate percentages of 10 to 20% for chronic mourning. It is unclear whether cases of delayed mourning are to be included in this number. However, since Bowlby bases his estimates on various studies in which delayed grief appears less frequently than chronic grief, an estimate of 15 – 25% seems to be a plausible range for the number of pathological mourners (Kleber & Brom, 1989).

7.4 Determinants of grief

The duration and the severity of a bereavement process depend on various factors. A series of factors determine whether a mourner becomes ill, whether the mourning reactions become chronic, or whether someone can build up his life again as it 'should be,' some of which have already been touched upon in previous chapters. Although they are dealt with extensively elsewhere in this book, it is worthwhile to mention a few of them briefly once more, as they can clarify the course and the nature of the coping process.

The determinants of the bereavement process have not been studied very systematically. Researchers have reported long lists of relevant variables, but the findings are not always consistent. In any case, it has become evident that the development of a bereavement process is influenced by a diversity of factors.

One important determinant appears to be whether the loss was expected. All 23 pathological mourners who were studied by Volkan (1970) were individuals who had lost their partner suddenly. The syndrome of unexpected grief was discussed in the previous section. Anticipation plays a central role in every traumatic event, as was mentioned in previous chapters, and is dealt with in chapter 9.

Other determinants refer to personality characteristics. Known personality traits (such as introversion/extraversion) are much less related to disturbed bereavement processes; more important are dispositions developed during childhood to form a certain type of relationship. The origins of these tendencies will be discussed in chapter 9.

Social factors concern the support of other people during the bereavement process and the rules and rituals that exist for mourning in a given society. Some cultures have extensive rituals and customs that allow the grief reactions to be 'channeled and removed', while other cultures, particularly in the Western world, are especially scarce in social rituals and customs (chapter 9).

The last determinant for grief we will mention here is the nature of the relationship with the deceased person. A loss is, without a doubt, much harder to bear and much more difficult to cope with for someone who had a very close bond with the deceased. A disturbed process of coping is rare when it concerns the loss of a friend or a distant relative seldom seen (Bowlby, 1980). However, the relationship need not necessarily have been a positive one. As outlined in the section on pathological grief, there exists a connection between ambivalence in the attachment between partners and bereavement problems; on the one hand there was affection, on the other hand irritation. It also became evident that the reactions were prolonged when the survivor had been dependent upon the deceased. In short, the psychological and social reactions to a loss are also determined by the strength of the bond with the deceased, the security of the bond, the ambivalence in the relationship, and the dependence on as well as the trust in the deceased for the execution of all kinds of tasks and roles.

The impact of the nature of the relationship with the deceased upon the bereavement reactions is clearly shown by the general finding that the death of one's own child is the most devastating loss (Bowlby, 1980; Gorer, 1965). The data of our outcome study of brief therapy (chapter 14) also points in this direction. Fifty percent of the persons treated for disturbances in the process of coping with a loss, were coping with the death of their child. In particular the death of a child caused by Sudden Infant Death Syndrome (SIDS) is a totally unforeseen and fateful loss (Beal, 1979; Krein, 1979).

7.5 Conclusion

Grief is a complex but coherent pattern of psychological, social and physical reactions to a clearly defined event; namely the loss of a loved one. Typical for grief are the following reactions: the search for (and 'finding' of) the deceased, anger and irritation, grief, the constant recollection of the deceased and his death, the evasion of memories of the loved one (though usually not a complete denial), feelings of guilt, physical reactions (insomnia, feelings of restlessness, etc.) and sometimes symptoms of identification with the deceased. Grief is not a process which takes place in a void. It has quite a few social implications. Expectations, role patterns and obligations are all changed by a loss.

In the pattern of reactions after loss we can identify a temporal development. The model presented was largely based on the theoretical approach by Bowlby and Parkes. Their perspective is also useful for the analysis of the coping process and effects of the loss of a partner due to divorce (Bloom, White & Asher, 1978; Weiss, 1976) and in other loss situations. A disadvantage, however, is that it has not yet been empirically substantiated by quantitative research with adequate data analysis. The studies of Parkes, on the other hand, excel because of their longitudinal design and their random samples.

There is a clear difference between the process of coping with a loss and the effects of the traumatic events mentioned in earlier chapters. It is true that in both cases there exists an experience of powerlessness, disruption and discomfort, and that many of the psychological processes and reactions correspond with one another. However, this similarity is not complete. In grief, the adaptation to a situation without a loved one plays an important role. A partner, confidant, and source of support has fallen away. The environment has changed drastically. The effects and the coping process are less focused on the event of the death, but more on the loss of the loved one and the building of a new life. Sadness and depression are therefore the emotions more strongly in the foreground after a loss, while anxieties play a less important role than in other traumatic events.

This is possibly the reason why bereavement problems are not included in the diagnostic category of posttraumatic stress disorder (chapter 2). Yet, this exclusion is criticized with increasing frequency (Sonnenberg, 1982). Indeed the death of a loved one is a 'stressor, which is exceptionally stressful for almost everyone'.

In other traumatic events there is always a loss present, perhaps not of important persons, but of ideas and views about oneself and the world that one is attached to (Janoff-Bulman, 1989). The existing order has been disrupted; a new order has to be built. Traumatic experiences always imply loss. 'It's as if my inside had been torn out and left a horrible wound there' (Parkes, 1972, p. 97). The mourner tries to understand what has happened. He continually tries to reconstruct reality. His ideas about himself and the world have become uncertain because of the event, and it takes time to adapt to the changes. These aspects are elaborated in chapter 8.

Grief is a meaningful reaction pattern, i.e. the functioning of the person may be unusual relative to other situations, but it is normal that someone shows the reactions and developments mentioned above. Grief concerns a process that enables someone to accept the loss in order to function again socially; otherwise problems arise. This is a very important conclusion that can be generalized to other traumatic events. One way or the other, a person must cope in order to function. The effects are expressions of a normal process, which generally cannot be qualified as unhealthy.

PART III

General Theory

CHAPTER 8

COPING WITH TRAUMA

The psychological consequences of various types of traumatic events were reviewed in chapters 3 through 7. It has become evident that extreme events may lead to considerable, sometimes lengthy, changes in adjustment and health. At the same time, however, disorders or prolonged consequences are not present in everyone, and reactions differ greatly depending upon the particular event and the individual.

In this chapter we offer a theoretical approach to the process of coping with a traumatic event and its implications. This approach is based on modern developments in stress theory, cognitive psychology and psychodynamic thought. It emphasizes the way an individual assigns meaning to and molds his situation. In this chapter, the attention shifts from the question of 'what causes stress?' to 'how does the individual deal with extreme stress?' The concept of coping is defined as a form of problem solving whereby the individual well-being is at stake and the person cannot fall back upon existing (routine) skills and options (Lazarus, 1981). 'Coping' is a broad umbrella term which relates to both situation-oriented behavior and to internal processes aimed at dealing with problems.

8.1 An integrative model of variables

A traumatic event leads to different kinds of reactions. How should envision this relationship between the event and its effects? We will present three perspectives on the connection between psychological and social phenomena (stressors) on the one hand, and illnesses and emotional disturbances on the other hand.

An implicit assumption exists, both in common-sense thought and in academic publications, that a traumatic experience leads to disorders. According to this perspective, which we refer to as the *traditional specificity perspective,* a certain psychological variable results in a specific health problem. In particular, psychosomatic literature from the forties and fifties, deals with this perspective. According to Alexander (1950), for instance, the continual suppression of anger leads to heart problems. Specific psychological characteristics are linked to different afflictions.

The research design of much social scientific research implies this notion of specificity. Only the relationship between one psychological variable and one health problem is examined. Whether other variables or illnesses play a role cannot be established on the basis of such a research design. Thus, in fact, a perspective of monocausality is presented.

This perspective is often used with regard to traumatic events, although it is seldom explicitly mentioned. A disaster is so extreme that everyone will exhibit disturbed behavior after the event. Dohrenwend and Dohrenwend (1981) refer to this as the 'catastrophe model'. In chapter 6 we indicated that many authors presume that captivity in a concentration camp will sooner or later result in a camp syndrome in everyone.

The inability to find specific psychological factors for every psychosomatic illness was partly responsible for the development of the *general susceptibility* perspective. A psychosocial variable – an event, a circumstance, a personality trait – does not specifically relate to one illness. Thus it does not cause a specific illness, in the way that a virus or bacteria causes a certain infectious disease. It increases the vulnerability of the individual, making him more susceptible to any disease.

This concept of general susceptibility is recognized in the 'stressful life events' approach, an approach in stress research mentioned in chapter 1. Time and again it has been shown in many studies that the number of extreme events in personal life, such as marriage, moving house, change of job, and being fired, are related to various health problems such as leukemia, depression, colds and suicides. The general thought behind this approach was that the more adjustment life events require, the lower the physical and mental resistance and the higher the probability of disturbances, whatever their nature (Holmes & Rahe, 1967).

The general susceptibility perspective also appears, for instance, in the research on the influence of traumatic events on physical health. For instance, the death of a partner has found to lead to an increase in the death and illness rates of widows and widowers as compared to those still married (Stroebe, Stroebe, Gergen & Gergen, 1981/1982). A stress situation demands so much from the organism that the least resilient part collapses. While in the traditional specificity perspective a traumatic event leads to a specific pattern of illness, according to the general susceptibility perspective, the event is related to various illnesses. The more traumatic the experience, the higher the number of unpleasant consequences. Both perspectives correspond in that they assume a rather direct link between the stress factor and stress reactions. A catastrophe is so traumatic that everyone will exhibit disturbed behavior, whether specific or nonspecific.

There are drawbacks to both perspectives, however. On a statistical level alone, there are shortcomings. The correlations between, for example, the number of stressful life events and the severity of the effects, though statistically significant, are low: approximately .10 to .25 (Rabkin & Struening, 1976). Statistically, the consequences correlate with various psychological factors as well. In short, the theoretical drawbacks imply that the traditional specificity perspective does not take into account the multiple determination of the phenomena to be studied, and the general susceptibility perspective does not explain the origin of a certain illness or disorder (also see Kleber, 1982).

We want to emphasize that in rejecting the linearity of the association between the stress factor and stress reactions, we do not allege that this connection is absent. On the contrary, Sherif and Sherif (1969) argued that one of the fundamental principles of human behavior is that 'structured stimulus situations set limits to alternatives in psychological patterning' (1969, p. 30). The more intense and inescapable the external situation, the less individual variation occurs in the experiences and reactions of those involved. What we are concerned with is that this idea is not sufficient for understanding the implications of traumatic experiences. Not everyone reacts to a situation in the same manner. The above-mentioned perspectives need to be supplemented.

A highly relevant psychological variable that is not included in research based on the above mentioned perspectives is the attribution of meaning. People perceive an event in a certain way, as a threat or a burden, but also as the end of a problematic state. Consciously or unconsciously, people interpret the circumstances. Soldiers who were admitted to hospitals because they had been heavily wounded in the war were pleased because they were escaping the even worse misery of the battlefield. Civilians with similar wounds in peacetime were melancholic and anxious (Beecher, 1956). Thus, the event acquires a meaning that can differ substantially according to the individual and the group. This aspect should therefore be incorporated into a model.

Figure 8.1 Stress model I

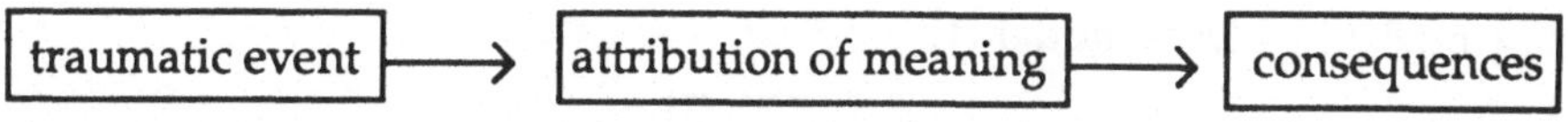

The event, which is experienced in a certain way, does not take place in a void, but in a context. It is not purely a flood that occurs, but an unexpected natural disaster resulting in many casualties. Such a disaster affects the family units and does not allow for individual activity. In short, various dimensions can be distinguished in a traumatic experience. Aspects such as the anticipation of the event substantially influence the way in which the shock itself is experienced and how people react. For instance, in chapter 7 we pointed to the effect of an unexpected loss.

Figure 8.2 Stress model II

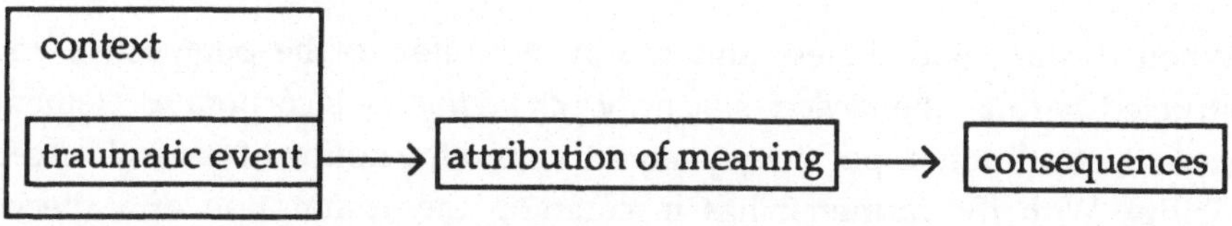

The relationship between event and consequences can be further elaborated upon. First of all, on an individual level, the characteristics of the person play a role. Age is an obvious characteristic but personality is also important. For instance, on the basis of certain earlier experiences, an attitude toward new situations has been developed that might be characterized by insecurity or an excessive need to control the situation. Such a disposition does not seem very conducive to handling overwhelming experiences (chapter 9).

The latter also applies to another determinant of the event-effect chain, namely the social network, i.e. the entire body of social contacts and relationships that surround and influence a person, and that guide the coping process in a certain direction. When someone can express his emotions to others, the effects of an extreme event prove to be less serious than when the person has to bottle up his feelings and has to cope with the shock in isolation (chapter 9).

When we add the above-mentioned aspects to the model, we obtain a markedly more complex picture of the relationship between event and effects.

Figure 8.3 Stress model III

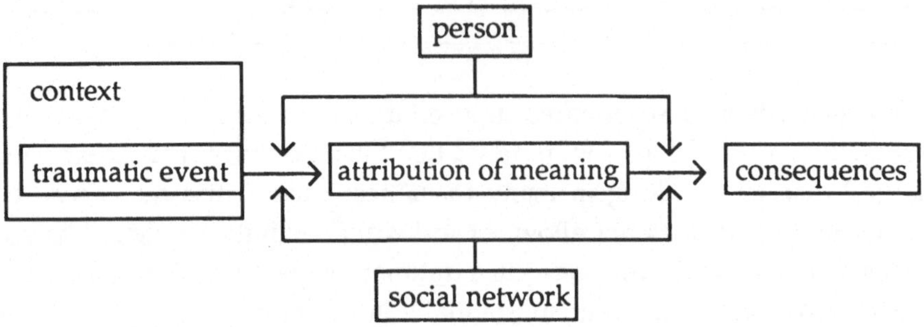

A final elaboration should be added, one that concerns the effects. Extreme stressors do not lead directly to health problems. A sequence must be distinguished in the effects.

When dealing with illness and health, we refer to the perspective constructed here as the *modern specificity perspective.* It incorporates elements from the traditional specificity perspective and the notion of general susceptibility. With the former it has in common the assumption of a specific relationship between psychosocial variables and effects; with the latter it is in agreement rejecting a monocausal explanation for symptoms. The perception of the stressor, as well as the nature of the stress reactions, personal factors and social relationships, determine the ultimate effects of an extreme event. Divergent psychological and somatic reactions are the result of divergent perceptions, emotions and responses. This perspective is pre-eminently a plea for multivariate studies emphasizing various interacting variables.

The model we have presented here, as a form of this modern specificity perspective, has not been elaborated upon much with regard to extreme events, but we regularly encounter it in modern stress research (Vingerhoets & Marcelissen, 1988). A well-known example is the so-called Michigan model, which was developed in order to study stress related problems in work. This model, which has led to many empirical studies, consists of a chain of variables; the objective environment influences the perceived situation, which in turn determines individual stress reactions with, as an ultimate consequence, illnesses, while personality and social relationships function as moderating variables (French & Kahn, 1962; Winnubst, Buunk & Marcelissen, 1988).

Numerous models have been formulated in stress research in order to illuminate the links between all kinds of psychosocial variables and health problems. The models are often placed alongside or opposite each other, as if they were actual data (Van Dijkhuizen, 1980), even more so when they are represented as drawings with boxes and arrows. The danger of 'reification' is present: the pretension that the model exactly reflects reality. These differences, however, are a matter of perspective. When examining the models under a microscope, differences in denominations, arrows and categories are apparent, but it should be remembered they are no more than schematic representations of theories.

There is a distinct development in the different theoretical approaches towards a common perspective on the relationship between circumstances and individual reactions. The 'stressful life events' approach mentioned earlier originally only attached importance to the effect of all life events experienced over a certain period, but later it evolved into a more differentiated theory (for example see Rahe & Arthur, 1978), in which there is room for intermediating cognitive, physiological and behavioral variables. The perspective of general susceptibility is thus replaced by a modern specificity approach.

In short, a model of relevant variables is useful in the sense of structuring and illuminating observations. It is, however, only a model and it remains just that. The time factor in particular renders the relationships in the model considerably more complex (time is difficult to represent). The environment influences perception, resulting in certain reactions, but in turn the reactions are interpreted and a change in attribution of meaning can alter the situation. A feeling of anxiety may result from circumstances, but may also lead to another interpretation of them. The direction of the arrows in the model can also be reversed. The same applies to the personality; it influences the outcome of an event, and this, in turn, may bring about personality changes.

We now focus attention on the coping process, the second question raised in the introduction of this chapter. First, we will elaborate upon the attribution of meaning.

8.2 Appraisal and meaning attribution

The attribution of meaning is an important element in current views of stress, trauma, and health. People interpret their circumstances in a certain manner and this interpretation strongly determines the ultimate outcome of a situation.

Richard Lazarus (1966, 1981) has analyzed the role of meaning attribution with regard to tensions, problems and ill health. He and his colleagues (Speisman, Lazarus, Mordkoff & Davidson, 1964), screened a shocking film showing a rite of passage among Australian aborigines for three groups of volunteers. The accompanying soundtrack was different each time. In the first group, the pain and the damaging effects of the rites were emphasized. In the second group, the effects were denied; the initiated were supposedly happy and elated about their initiation. The third group heard a detached commentary from a scientific perspective. During and after the film, the researchers determined different physiological processes and asked the viewers for their opinions. It was established that the latter two groups exhibited substantially fewer stress reactions (anxieties and various physical phenomena). The commentary made it possible for them to view the situation as less threatening.

Lazarus employs the term 'appraisal' in this context. It refers to a person's 'evaluation' of the situation. An individual immediately and automatically interprets everything he encounters in terms of the relevance and importance to himself. Perceiving always involves interpreting and judging at the same time, however implicitly. 'Appraisal' is not equal to a conscious evaluation of a situation. The term refers to a quick evaluation of the value of a situation without much reflection. In sports such as table tennis and badminton, a player 'knows' how to hit the ball in order to make it more difficult for the opponent to strike back. In a flash, he takes into account the velocity of the ball, the position of the opponent and a number of other details. This quick decision certainly is not a reflex (a learning process has preceded it), but neither is it a conscious evaluation of the situation. The term 'appraisal' refers to this direct, almost automatic interpretation of the situation. It is a wordless awareness of the situation, as it were. And it is this interpretation that Lazarus manipulated in the above-mentioned experiment.

The role of the attribution of meaning becomes clear in the first reaction to a shocking event. There are well-known stories about people who almost automatically save other people's lives and their own, only to faint or start crying afterward. This is not the behavior of a machine, however. The person interprets the situation with incredible speed and reacts on the basis of this 'appraisal'. The phenomenon of acute depersonalization, mentioned in the earlier chapters, is an indication of this (Lazarus, 1981).

Lazarus distinguishes three types of appraisal (Lazarus & Launier, 1978). The first is primary appraisal: the evaluation of the situation in terms of whether it creates problems for personal well-being. The interpretation can imply that the situation is judged to be: 1) irrelevant, 2) positive, or 3) stressful. In the latter case, there are once again three forms of interpretation: harm, threat or challenge. Harm refers to damage already done, such as the loss of a life partner. Threat refers to something that has not happened yet, but which is imminent. The anticipation of an unpleasant event may just as easily lead to stress reactions as the event itself. Finally, challenge refers to the possibility of growing into a situation.

This primary process of interpretation determines the intensity and quality of the emotional response to the interaction between the person and the situation. A positive appraisal results in pleasurable emotional reactions such as joy and relief. The interpretation of a situation as stressful leads to negative emotions, such as anxiety, fear and guilt. For each emotion, there is a specific pattern of interpretation. 'Anxiety, for example, involves an anticipated harm (threat) that is ambiguous, either as to what is to happen or what can be done about it' (Lazarus, 1981, p. 193).

When a situation is perceived as stressful, secondary appraisal follows. The individual now asks himself: what can I do about it? He can escape, he can deny the threat or he can try to change the circumstances. There are many ways to deal with a problem. In order to select an adequate way of handling the stressful situation ('coping') the person uses information about his own skills, the demands of the situation and the usefulness of the various ways of dealing with the circumstances concerned. 'Whether consciously and deliberately or unconsciously and automatically, a decision is made about what to do' (Lazarus, 1981, p. 195). How someone reacts in a situation perceived to be threatening or painful depends upon this interpretation of the options for coping with the problems involved.

Both appraisal processes are connected with one another and determine one another. The skills and means that someone believes he possesses, influence the perception of threatening situations and unpleasant problems. Both forms of appraisal refer to a focus on, and evaluation of, information in a situation: does the situation imply problems and, if so, which options can be used? Lazarus strongly emphasizes the complementary character of both forms of interpretation, and later even regretted the use of the terms 'primary' and 'secondary' (Lazarus, 1984). Both processes are so closely linked and alternate so often, that the suggestion of an order is incorrect.

On the basis of information about his own reactions and those from the environment, a person interprets his circumstances anew. The relationship between person and environment is, after all, constantly in motion. The term 'reappraisal' is employed for this re-interpretation.

Lazarus' ideas clarify the position that meaning attribution maintains in the adjustment to severe changes. In more dated literature, it was often too easily assumed that what is stressful for one person is also inevitably stressful for the other, and that behavior would be identical. The differences in appraisal lead to the fact that people – even in extreme circumstances – cope with the event in very different ways.

In the perspective described above, emotions are considered a result of thought processes, which in turn are set into motion by events. 'The nucleus of an emotion consists of a change in behavioral tendency, provoked by a situation as perceived by the subject' (Frijda, 1982, p. 246).

This perspective shows a strong resemblance to cognitive views in experimental and social psychology (Lundh, 1983). According to Neisser (1976), cognition refers to processes in which sensory input is transformed, reduced, processed, stored, recalled and used, even in the absence of relevant stimuli. All sensory input is subjected to a quick, mostly unconscious analysis of meaning, i.e. a structure is sought in which the input can be placed and which gives it meaning. As mentioned earlier, this does not have to occur on a conscious level. A person usually registers meanings without consciously dwelling on them. Only when an event is so complex that its meaning cannot be determined immediately, does a more conscious analysis of meaning take place.

This train of thought might suggest that emotions are a negligible by-product of cognitions; this is not correct, however. The only concern here is emphasizing that they are *linked* to an interpretation of a situation. Cognitions or meanings constitute necessary conditions for emotion, but once these emotions are evoked, they in turn influence the various appraisals (Lazarus, 1984).

The specific links between cognitions and emotions may be clarified with an example. Anger is a consequence of the awareness that someone else is being held responsible for an event and that it is still possible to change the situation. This was clearly expressed in Bowlby's attachment theory. After separation the individual remaining behind protests against the loss. He still holds the idea that he can exert influence on the situation, which distinguishes it from the example of sadness, an emotion that is linked to the awareness that something has been lost and that this loss is irreparable. This linking of emotions to specific cognitions will not be elaborated upon here (we refer to the analyses of Averill, 1983, Frijda, 1986, and Smith & Ellsworth, 1985).

The perspective mentioned above makes it understandable that emotions can change rapidly. A person often experiences various rapidly changing emotions, sometimes even simultaneously (Folkman & Lazarus, 1985). It was shown in chapter 3 through 7 that various symptoms occur simultaneously after traumatic events. After all, more than one meaning can always be perceived in a situation. The awareness of what has happened also changes continuously, thus the emotions shift along with the changing appraisals.

The concepts dealt with in this section clarify the processes that occur when one finds oneself in a threatening and taxing situation. Lazarus has constructed a conceptual approach to the individual interpretation of this situation. This general view applies to all 'demands that tax the individual', although it is not very precise about the conditions under which certain appraisals may occur.

Lazarus' work makes it clear that the attribution of meaning is important, with regard to traumatic stress as well, but it does not explain why one is haunted by a traumatic event for such a long time. An action that is related to an event often follows an emotion, so that the cognitive processes are changed and the emotion decreases. This is not as simple in the case of a traumatic experience. After the event itself has been over for quite a while, the person still has to deal with the consequences. The approach mentioned here does not offer an explanation. We therefore turn to another analysis.

8.3 The concept of schema

The traumatic event has passed. Nevertheless, the victim still suffers. Even when there are no material and physical consequences, various psychological reactions often remain for a long time after the event has come to an end. Something continues on internally, as it were. A mental process in which the shock is 'digested' is necessary in order for the person to be able to function normally again.

Before elaborating on this process, we wish to present a concept that clarifies why the event remains manifest in behavior, thoughts and feelings. This concept, originating in cognitive psychology, plays a role in the cognitive theories to be outlined below.

Individuals have built up opinions and expectations about themselves and the world. Their existence proceeds by means of rules that have evolved in the course of a development, but that are rarely verbalized. All these rules and opinions are so self-evident that they are used without too much prior consideration. When I turn a switch, the light goes on. I am surprised when this does not happen. In my mind, the act and the expectation are stored together.

In cognitive psychology, such expectations are referred to as *schemata*, a concept which was introduced by Sir Henry Head in 1920, and which was further developed by Bartlett in 1932. The latter psychologist was dissatisfied with the simple perception theory of his time, and he used the term to explain that perception is organized and guided by expectations, which in turn can be changed by the consequences.

The schema is the central cognitive structure in perception. It belongs to the perceiver, can be changed through experience and is reasonably specific, according to Neisser (1976). The schema accepts information when it is available. It guides movements and activities. Man does not react blankly to information, but employs certain structures into which this information fits, which process the information and relay it.

The schema concept had already appeared in Piaget's early work. A child develops ways of thinking and cognitions during his development. At the beginning, a toy does not exist anymore for a baby once it is out of sight, but then gradually the child develops the concept of an object, independent from the direct perception. In the child's mind, a 'framework of thought' grows; objects continue to exist while they have disappeared from sight. Piaget refers to the framework of thought as a schema. It is a psychological structure for understanding the world. It should be evident that this is not a matter of conscious processes. The child cannot verbalize clearly, but begins to obtain 'knowledge' of the world. Mental pictures emerge about what can be done with various objects. This is the nonverbal notion, as was discussed in the section on appraisal.

In the course of the development, these schemata are formed and expanded upon. Piaget refers to the attunement of reality and the internal cognitive structure as adaptation, in which he distinguishes two complementary processes. Accommodation refers to the modification of schemata to fit new information. Assimilation implies the incorporation of information in already existing frameworks of thought. Roughly, it can be stated that assimilation implies the adaptation of the surroundings to the individual and accommodation implies the adaptation of the self to the surroundings.

Both processes can be discerned in individual behavior. People continuously absorb new information that fits in with existing beliefs, but at the same time they must change their ideas as well, so new information will fit in. This aspect will be referred to in the following section.

The schema concept is an 'umbrella' concept, which occupies an important position in cognitive psychology and in other psychological sub-disciplines as well. The concept is defined as a knowledge structure, stored in memory, that consists of elements of past experiences and reactions. Although it is an abstracted construct, it is based on concrete experience and it provides a basis for behavior (Landman & Manis, 1983).

The concept of schema is exceedingly broad and not very precise. Yet it is useful and guides thinking. In terms of these schemata, the earlier mentioned process of 'appraisal' can be understood. What is in particular important to our discussion is that the concept of schema refers to general constructs about the self and the world, which guide individual thought and action and help to comprehend the world (so called 'higher order schemata'). It is these schemata that fail after a traumatic event.

8.4 An analysis of the coping process

A woman loses her husband due to a heart attack. Suddenly, she is no longer a wife and she is no longer in a close relationship with the other person. Her feeling of security has vanished. The old expectations, notions and opinions about herself, her partner and the world no longer apply. The traumatic event has upset these schemata. A vast amount of information, to which the person cannot adequately react, disturbs the entire process of thinking, feeling and acting. In Freud's words: 'a trauma takes place when an excessive amount of stimulation overwhelms the emotional apparatus'.

New assumptions, expectations and ideas about oneself and about the world must be constructed. The coping process can be regarded as a process of change of old schemata towards new ones. Accommodation takes place. This idea was elaborated upon by Horowitz (1976, 1979), who has developed a very well-conceived and complex theory about coping with stress, on the basis of psychoanalysis and the stress approaches of Lazarus and Janis. Elements of this theory will be discussed in this section (see chapter 12).

Horowitz' theory can be summed up in a nutshell: 'negative stress stems from experience of loss or injury, psychological or material, real or fantasized. If action cannot alter the situation, the inner models or schemata must be revised so that they conform to the new reality' (1979, p. 244).

As mentioned, notions about ourselves, concepts, illusions, assumptions about social relationships and all types of other images, at least partly determine human experiences and activities. About the self-image, Horowitz says: 'A person has a dominant self-image as competent that is relatively stable and usually serves as the primary organizer of mental processes' (Horowitz, 1979, p. 246). Thus, the self-image is a structuring system, as is the schema mentioned in the previous section. A shift occurs from a competent self-image to a self-image which is characterized by feelings of worthlessness and unsuitability, for example, due to being fired and subsequently unemployed. The person feels useless and ashamed. Experiencing a crime of violence nullifies the usually implicit feeling of security that most people have. The sense of invulnerability shatters; in the victim's mind, a similar crime could occur any time again.

Extreme experiences are viewed against the background of notions about the self and the outside world. The new information – the event is regarded as information – and its concurrent associations are seen in terms of these inner models. 'Working through the news involves relating to it according to several models' (Horowitz, 1979, p. 264). Traumatic experiences thus create a change in individual opinions, ideas and expectations. New images are developed and the old ones acquire a new meaning.

People have a need to harmonize new information with the presuppositions and notions, based on earlier information. This need is referred to as the completion tendency. Horowitz' analysis of the completion tendency is a cognitive reformulation of Freud's repetition compulsion. The memories are presented time and time again, until the schemata agree with them. The events are reexperienced each time, until finally, the reality and the inner models are adapted to each other.

For Horowitz, an active memory storage results from this completion tendency. Normally, an event is perceived, coped with and sent on to the passive memory for storage; however, after a traumatic event, which implies intense change, the experience returns to active memory (which more or less equals short-term memory) and is then reexperienced again and again.

This active memory is characterized by repeated representation. The memories are presented over and over until they are processed. In this case, the coping process is completed when the particular content is no longer stored in active memory, but in passive memory. The danger of a tautology exists here; something is reexperienced because it is in the active memory, but the active memory in its turn is defined as the continual reexperiencing.

The degree to which the information keeps returning, is regulated by so-called 'controls', processes that govern mental functioning, such as the defense mechanisms formulated in psychoanalysis. We are dealing with an abstract and occasionally obscure construct, which Horowitz utilizes to refer to various regulatory processes. Control processes operate, among other things, by selecting information and the models – beliefs and images about the self and the relationships with other people – which regulate the interpretation of information. For instance, a victim of violence may come to feel more attached to other victims, or in Horowitz' terms: he handles his emotions by using schemata in which he feels more attached to fellow sufferers.

Concrete examples of control processes are: active inhibition of representations (i.e. the active warding off of the reexperiencing of aspects related to the traumatic event), the search for new information, the shift of meanings and the revision of schemata. These control processes work in such a way that the continuous representation of the event is inhibited or accelerated. Optimal control delays the intrusion and yields tolerable dosages of the new information and the emotions. This leads to an optimal alternation between denial and intrusion.

Too much control prevents the repeated representation in the active memory, denial becomes too strong ('numbing') and prevents further processing. Too little control gives way to excessive emotions and a continual return of the traumatic experience. An optimal control leads to both denial and intrusion. Thus, completion can occur; inner models can adapt to reality.

Control processes are a type of regulatory processes, which take up a central position in the system described above, in which information is regulated and emotions are reduced or activated. A simple representation is given in figure 8.4.

Figure 8.4 A simple representation of Horowitz' coping theory

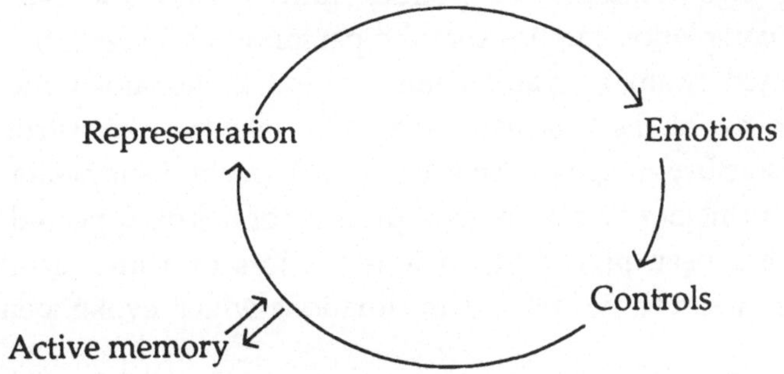

Horowitz follows the theory of Freud, who stated that the emotional apparatus defends itself against too many stimuli by means of a stimulus barrier. When someone experiences a threat, anxiety is activated. When anxiety increases, the stimulus barrier is raised, thus decreasing the input of stimulation. The situation is dynamic:

more input ⟶ more anxiety ⟶ more regulation (modulation) of the input
⟶ less input ⟶ less anxiety ⟶ less regulation ⟶ more input etc.

In this perspective, anxiety and the stimulus barrier are theoretical constructs and not direct emotions and realistic thresholds. We are dealing with heuristic concepts, although Horowitz' objection that the word stimulus barrier implies too concrete a metaphor is justified. He therefore replaces it with input modulation.

Horowitz' approach is based upon a welding together of this psychoanalytical interpretation with ideas from stress researchers Lazarus and Janis. Three general systems are formulated in this theory: the 'ideational system' (perception, representation and different cognitive processes), emotions and control processes. Following Lazarus, emotions are represented as reactions to thought processes (and not so much vice versa, as in traditional learning theory, where anxiety is often considered a primary motive). Emotions motivate 'controls' which in turn reduce and increase cognitive processes, thus creating an equilibrium.

When the anxiety, fear or anger is too great, these controlling processes see to it that the intrusion decreases. This leads to a decrease in emotions, resulting in a lower level of required control, so that representations and emotions can increase again. This is the characteristic alternation response – Horowitz speaks of 'oscillation' – between intrusion and denial. Alternating between those two tendencies continues until completion has been achieved. This completion implies that the person is no longer, or hardly ever, overwhelmed by anxieties and memories, and, at the same time, he no longer tends to avoid the images of that time. He has reconstructed a reasonably satisfactory image of himself and the world. Completion does not imply that the memory is no longer unpleasant or has disappeared. Even after an event has been processed, such as the loss of a life partner, for example, the grief sometimes returns in situations which evoke memories about the loss.

The result of the coping process is that new information agrees with existing expectations, ideas and strategies and that these schemata in turn are adapted to the information (assimilation and accommodation in Piaget's terminology). The ongoing spiral 'representation – emotion – control' (figure 8.4) has come to an end.

Summarizing, what are the important characteristics of the theory on stress response presented here? The human mind utilizes existing inner models to interpret new events: 'models based on the past must interpret the present and be revised to meet the future' (Horowitz, 1979, p. 235). An event of an extreme nature causes the balance between old knowledge and the acceptance of new information to be severely disrupted.

The result is a discrepancy between the implications of the event and the relevant schemata of the individual. This discrepancy creates emotions. Traumatic events are so divergent from the inner models that very painful emotions emerge, as a result of which control processes are activated to avert the threat of unbearable grief. The resulting coping process is schematically represented in figure 8.5.

Figure 8.5 The interaction between ideas, emotions and 'controls' (from Horowitz, 1979, p. 250)

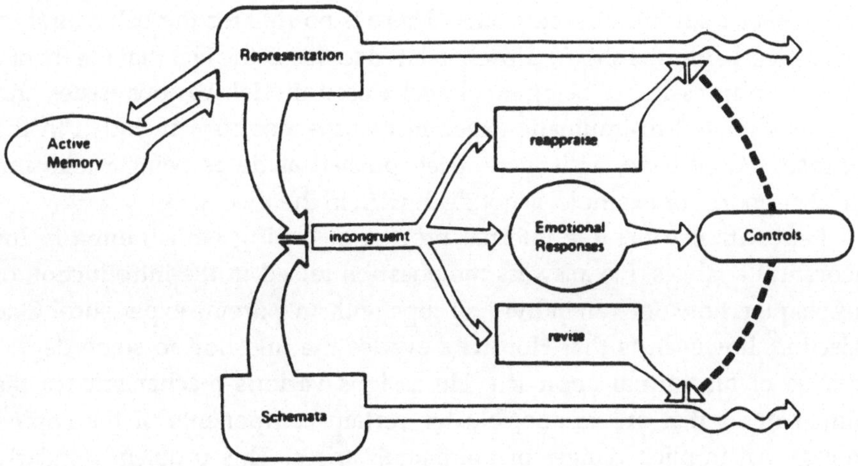

Normally, an event leads to emotional reactions, which, after perception and other cognitive processes, evoke an act that in turn changes the situation. Direct action eliminates the problems evoked by the event. This is not possible, however, in the case of a traumatic event. The act cannot eliminate the changes caused by the event. That is why the problems have to be solved on an intra-psychic level. Through and in these processes the person gets a grip on his situation after all.

All the events described in this book entail a disruption in existing schemata relating to the self and the environment. The individual expectations, opinions and ties to other people are partly lost. These must be replaced. Horowitz' theory deals with the painful process between the moment of the loss and the ultimate adaptation. During that period, intrusion and denial occur. The new information is gradually absorbed and accepted in the individual's thoughts, expectations and assumptions. This coping process is a slow one: 'time is essential for review of the implications of the news and available options for response' (Horowitz, 1979, p. 248).

Horowitz analyzed how someone re-establishes a level of stability and control from a state of powerlessness and bewilderment. The analysis is strongly cognitive; a term such as information processing is continually utilized. A concept such as defense mechanism is placed in a cognitive framework. The orientation is, in this context, directed to the inner (intra-psychic) processes and mechanisms after the event. This is partly caused by the fact that traumatic experiences, by definition, concern events in which action cannot change the situation.

However, in his formal analysis, Horowitz is barely interested in processes other than intra-psychic ones. There is no role for the behavioral or social effects after the shock. Closely related to this is the fact that the theory gives the impression of being an almost closed model. Various issues and questions related to traumatic experiences have not been included in the approach. They have, as it were, been pushed aside as being irrelevant. Social support, for example, is not dealt with in the theory.

Horowitz clarifies the internal process of coping with trauma in his theoretical analysis. He answers the question raised in the introduction of this chapter: how does an individual cope with an extreme experience? One objection, however, is that Horowitz evades the question to some degree because of his formal approach. He assigns various mechanisms to the human mind that are responsible for certain components of the coping process. An implicit danger of homunculi arises. This problem regularly emerges in psychology. The memory theories, in which the human memory is explained in terms of mechanisms which in turn possess a memory as well, are well known. Sometimes Horowitz explains human reactions to a shock with the aid of mechanisms and processes which, in turn, react in same manner. This is especially true of control mechanisms. They stimulate or inhibit the input of information. Regulating bodies evaluate the severity of the emotional reactions and select images and schemata. In short, they function as small scale copies of the human mind.

However, the homunculus problem is characteristic for many modern, cognitive approaches (Dennett, 1980). Horowitz' theory explains the dynamics of the process of coping with traumatic stress very well. This process is analyzed in significant delineated mechanisms and processes. That is why the theory has a clear heuristic value. The fact that these processes and mechanisms, which are explained on a general level, are then postulated once more on a specific level, is a problem which cannot always be avoided. The question of the homunculus in the formation of psychological theory is a problem which is very difficult to solve. It is quite an achievement when it is known where exactly a homunculus is hidden in a theory (Wilkins, 1981). This does indicate, however, that a further elaboration of the obscure concept of 'controls' would greatly benefit Horowitz' theory.

8.5 Meaning and control

Horowitz' analysis of the process of coping is a formal approach. It indicates the mechanisms with which the traumatic event is processed internally. Less attention is devoted to the *contents* of the coping process. What kind of new information is produced by traumatic events and why do existing schemata need to be aligned with this new information? Are there discernible patterns in this information? Or are there specific schemata which are likely to be disrupted by severely distressing events?

We place three closely related propositions centrally in this section:

1) After a traumatic situation, the person searches for the meaning of what has happened to him;
2) The person attempts to regain control of the situation;
3) Social-psychological processes play an important role in the effects and the process of coping with a shock.

These propositions can be clarified by theoretical and empirical findings of several studies in the fields of social psychology, clinical psychology and stress research. First, the orientation of these studies is explicitly cognitive: the focus is on the ideas, assumptions, expectations and explanations utilized by individuals. Cognitive elements are supposed to determine behavior, experience and possible mental disorders. A human being behaves in accordance with his (often subconscious) beliefs and assumptions.

The approach we present in this section is a social psychological one. The cognitive aspects are placed in a social context. The presence of others and their beliefs co-determine behavior and experience of the individual. It is not surprising that these studies were strongly inspired by the approach of Lazarus, by cognitive innovations in learning theory (Peterson & Seligman, 1983), and by certain social psychological theories, especially social comparison and attribution theory. Taylor (1983), who studied the coping process of women suffering from breast cancer, is a representative of this approach. The ideas on control are further elaborated upon by Rothbaum, Weisz and Snyder (1982). There is no neatly defined theory but rather a heterogeneous group of researchers who developed a number of theoretical thoughts and applied them to the study of stress phenomena.

A distinctive notion is that when confronted with dramatic circumstances, people react with various, mostly cognitive, attempts to regain their original well-being. Two general themes can be distinguished: the search for meaning and an attempt to get a grip on the experience and its implications. We discuss these themes utilizing the theoretical and empirical findings of the researchers mentioned above.

The search for meaning

Usually, people's perceived worlds are reasonably orderly and comprehensible. The individual knows what to expect and why a certain event occurs. There are certainties expressed by expectations, presuppositions and ideas which guide behavior and experience. A traumatic event destroys this secure world. The order and regularity suddenly vanish. In chapter 1, we pointed out this acute disruption of 'normal' existence.

The world of certainties – 'the assumptive world' (Janoff-Bulman & Frieze, 1983) – must be reconstructed. People strive to reconcile old schemata with new information through a search for meaning.

A victim tries to find the meaning of his existence after a traumatic event because its order has been disrupted in an initially incomprehensible way. He poses the question: 'why me?'. The problem of loss of meaning often seems to focus not on the question 'why did this event happen?', but rather on the question 'why did this event happen to me?' (Janoff-Bulman & Frieze, 1983, p. 6). People know quite well that a traffic accident could happen to them, but why does it actually happen to them?

The search for meaning has been investigated explicitly in a number of studies. Although these studies did not deal with events that are central in this book, the results are comparable. Taylor (1983) studied women who had discovered they had breast cancer and underwent an operation. The search for the reason for what had happened to them proved to be a central theme in the various research findings. This expressed itself, among other things, in the search for and identification of an explanation for the disease.

Nearly all patients reported 'clear causes' for their breast cancer, while, in fact, the etiology of this disease is quite uncertain from a medical point of view. The pill, divorce, stress, wrong food, air pollution, etc. were given as causes. The importance to the patient of finding an explanation is also evident in the fact that the husbands in question reported significantly fewer causes than their wives. On the basis of questionnaire data and the general practitioner's judgment, Taylor concluded that the specific content of the explanations given, did not relate to the woman's well-being. This means that people search for a causal meaning behind traumatic experiences, and that its nature is of secondary importance. What matters is actually finding an answer.

The destruction of the notion of living in an orderly and meaningful world and the search for meaning were also examined in a study on 77 women who were incest victims (Silver, Boon & Stones, 1983). Although the events had occurred years ago, the search for understanding still continued. The more active the search, the more there occurred ideas, outbursts and nightmares relating to what had happened. The quest for a meaning created painful emotions and displeasure, which in turn evoked 'the search for meaning'. This is, once again, evidence for the idea, formulated by Freud (1920), that reexperiencing is related to the search for meaning.

Did these women find a reason for what had happened to them? Fifty percent did not succeed, even after 20 years (Silver et al., 1983). Social support appeared to be the most important factor in finding or not finding a meaning. It was beneficial if there was someone to whom the woman could express her feelings about incest.

The women who could find a meaning, usually realized this through a clarification of the circumstances of what had happened. They reported poor sexual relations between the parents, the motives of the father and distorted family ties. An important finding is that one specific explanation was no more desirable for one's well-being than the other.

With regard to the search for an explanation, attribution theory (already mentioned in the chapter on violence) may be used to explain research findings. When someone is confronted with a sudden change in situation, he searches for the cause for that change. In that way he can explain, predict and control the threat. Consequently, he *attributes* the event to a certain cause.

In their research of a group of victims of severe accidents, Bulman and Wortman (1977) found that self-blame was related to better functioning.

Blaming oneself for the problems enabled the person to continue to believe in a world in which outcomes are meaningful and controllable, because the victim can minimize the likelihood of future occurrence of such negative outcomes. However, this finding has not always been replicated in later studies (for instance, Taylor et al., 1984). The term self-blame is somewhat confusing. It seems paradoxical that a belief that a negative event was intentionally brought about can lead to positive adjustment. The concepts causality, responsibility, and blameworthiness are not interchangeable (Shaver & Drown, 1986). The term self-attribution of causality would be more appropriate.

From further research on the women with breast cancer (Taylor, Lichtman & Wood, 1984), it was found that the women, contrary to what might be expected, did not look for a cause immediately after the event. It was not until later in the coping process that this became an important issue.

Taylor and her colleagues also came to the conclusion that the contribution of attributions to understanding the process of coping with cancer, was limited. Although people searched for a cause for what had happened to them, the result of this attribution process was not very important to an adequate coping process. Taylor et al.'s statement on this matter is relevant: 'Efforts to find meaning in the cancer experience seemed to be reflected more in respondents' changes in their attitudes towards life than in their attributional explanations for cancer' (1984, p. 499).

What matters most is that people find an acceptable explanation of what has happened. What does my life mean after the event? What are the implications of the event for my life? Trying to find a cause for a traumatic event is part of a widely oriented process of searching for meaning (Van den Bout, Van Son-Schoones, Schipper & Groffen, 1988). Women from the breast cancer study appeared to re-interpret their life. After the illness became known, they started to live more on a day-by-day basis, to develop themselves more and to think more about themselves in general.

In short, the lives of these women were restructured. Family relations and personal activities became more important than external obligations, for instance. In terms of the coping theory discussed above, it could be said that the existing schemata had been changed through new information. According to Taylor (1983) the threat of cancer had been experienced as a catalyst towards restructuring their lives in a more meaningful way.

We have encountered this re-interpretation in earlier chapters, especially in the chapter on grief. A traumatic event is a breach in the course of life. The afflicted person must develop new rules, expectations and norms for himself. He searches for a meaning. In a traumatic event, this quest is almost always related to death. A person has been confronted with the threat of his own death or with the death of another. This confrontation creates anxiety and pain; it undermines the notion of invulnerability and makes a person aware of his own mortality. The study of traumatic reactions is the psychology of the survivor, according to Lifton (1979), who places this contact with death in a central context. His frequent use of the term 'death imprint' is evidence of this.

The process of coping with this confrontation is reformulating a new or renewed sense of one's own existence. An existential – and for Lifton (inspired by phenomenological philosophy) even a moral – dimension is thus linked to the search for meaning. A new perspective on life must be formulated by the victim or survivor. The same is argued by Frankl (1959) when he states that the quest for meaning – the constant search for the understanding of life events – is the central motivation of the human individual.

The regaining of control

The more control people can exercise on a situation, the less they suffer from diverging symptoms and the better they will function in general. This is the general conclusion of the many studies, that focus on the phenomenon of control (e.g. Rothbaum et al., 1982; Weisz, Rothbaum & Blackburn, 1984), as well as various studies on traumatic experiences. In earlier chapters, we stressed the importance of psychological control over a situation. In chapter 1, we described a traumatic experience as being pre-eminently a confrontation with an intense powerlessness. The affected persons attempt to regain their grip on the situation after the event has occurred. This is pure necessity if they are to function adequately again. In this section we further analyze this search for control.

Lack of control

Seligman arrived at his classic theory about 'learned helplessness' in the 1960s by conducting experiments on animals. After dogs had been exposed to unavoidable electric shocks, they acquired a passive and helpless attitude, which continued in later situations, even when they could escape from the shock by running into another kennel, for instance. A striking example is a study with rats (Seligman, 1975). When placed in a large barrel of water, from which they could not escape, rats sometimes swam for 24 hours before giving up. Rats, who had been taught helpless behavior in previous situations of uncontrollability, drowned after only a few minutes.

Afterwards, Seligman (1975) used this concept of learned helplessness to explain human behavior, especially depression. Confrontations with uncontrollable circumstances cause a pattern of helplessness, loss of interest, emotional numbness, anxieties, etc. This pattern is generalized to other situations than the original ones. Learned helplessness becomes an expectation of 'response outcome dependence' which crops up in all types of new situations.

Seligman's work illustrates the break of the modern learning theory with the traditional learning psychology. A cognitive element, namely the expectation that a reaction will or will not lead to a result, is introduced in order to understand behavior (although Tolman had stressed the central place of expectations as far back as in the thirties).

According to Peterson and Seligman (1983), from this pattern of learned helplessness, a diversity of passive and other so-called maladaptive reactions to traumatic situations can also be understood. An uncontrollable, unpleasant stressor occurs, and a general expectation of future powerlessness arises. The affected person begins to view himself as a victim who can be hurt time and time again. In addition, a series of complaints and problems emerges in situations that are not directly linked to the original stressful situation.

Caution should be observed in applying the model of learned helplessness to traumatic stress. A traumatic experience does not only lead to apathy, numbness and other expressions of passivity. In addition, posttraumatic pathology is not always manifested in depression (chapter 7); passivity and helplessness are not always the most striking symptoms. Sometimes overalertness and startle reactions prevail. Also learned helplessness is a pattern that is learned mostly through many 'trials', while a number of the traumatic events dealt with in this study are one time events.

It is important to realize that uncontrollability does not necessarily lead to helplessness. This depends on the interpretation of the situation. When the person blames his inability to control the situation on his own general shortcomings – 'I am a failure' – helplessness occurs earlier than when he blames it on accidental circumstances.

This is why Seligman later added an explanation, in terms of attribution theory, to the theory described above. When confronted with a situation, a person asks himself 'why?' and the answer to this question strongly determines his reaction. Three general dimensions of causal explanations can be distinguished: internal – external (the cause lies with the person or in the circumstances), stable – temporary (the cause is permanent or temporary), global – specific (the cause permeates all kinds of aspects or only one). When the person interprets the cause of an event as internal, stable and global, then reactions of helplessness occur most readily (chapter 11).

The search for control

The effects of the lack of control in an extreme event are only one side of the picture. As we stated in the first section of this chapter, we can outline a model in which situations lead to certain reactions. The direction of this relationship can also be reversed: certain reactions change the situation. Helplessness is a result of traumatic events, but, at the same time, people attempt to regain their grip on the situation, and to battle this feeling of helplessness.

We can observe this quest for control in the process of coping with a shock. Taylor (1983), in her study on cancer patients, even calls it one of the most important themes of the coping process. Many of the women interviewed were of the opinion that they could exert control over the course of their disease. This belief in their own control was clearly stronger than relatives' and friends' impressions of the patient's control of the disease. Moreover, it was shown that well-being was significantly related to the belief in one's own control and in the doctor's control over the disease.

Many of their attempts to regain a grip on the situation were psychological: 'I think that if you feel you are in control of it, you can control it up to a point' (Taylor, 1983, p. 1163). The assumption of a discontinuity between the period before the occurrence of cancer and their present life was characteristic. They believed that things were different now. The causes responsible for cancer had been removed, and they were consequently less vulnerable. In this way, the women created a certain sense of control.

In an interesting follow-up study, Taylor et al. (1984) compared four different forms of control methods. So-called cognitive control – starting to think differently about the traumatic event and its implications – was clearly more conducive to the individual functioning and well-being, than other forms of attempted control.

The behavioral form – changing the circumstances through direct action (changes in diet, for example) – was less influential. This also applied for the informative form of control practice, e.g. acquiring information about the causes of cancer. In the discussion on attribution theory, it was shown that this form of coping with a shock is not necessarily effective.

The fourth form of control which Taylor examined was the retrospective practice of control. In retrospect, the person in question comes to the conclusion that he has dealt with the traumatic situation and will succeed in doing so in the future. This retrospection, in a way a type of cognitive control, was not at all related to the individual functioning and well-being. The latter fact actually contradicts Seligman's idea about helplessness as a consequence of lack of control and indicates the importance of a self-generated awareness of control.

Thompson's (1981) review also showed that cognitive control was the most successful factor in reducing stress reactions. The effectiveness of all stress coping, i.e. all forms of control, is strongly dependent on the situation (Kleber, 1982).

Vulnerability

The extreme powerlessness, experienced as a consequence of a traumatic event, implies a confrontation with one's own vulnerability (Janoff-Bulman, 1989). Therefore we discuss this aspect in this section. The striving towards regaining a grip on the situation can be seen as an attempt to deal with the idea, obliterated by the shock, that one is the master of one's own existence.

In daily life, people implicitly feel they will be safeguarded from disasters: 'it won't happen to me'. This awareness is functional, because it prevents someone from dwelling incessantly on the dangers of daily life, such as crime, illness and accidents – although it does entail some risks (an obvious example is the persistence of smoking, regardless of all information about associated dangers).

A human being has a sense of invulnerability. Perloff (1983) gives some examples from scientific research in her review; almost 90% of American drivers feel that their driving is above average; on the one hand, people are afraid of cancer, while on the other hand, they largely underestimate their own chances of dying from this disease; on the average, interviewed persons assessed their life span to be 10 years longer than the life span of the population. Time and time again, it has been shown in research that people systematically underestimate their own risk of failure or of becoming ill in comparison with others.

The source of this feeling of being invulnerable may be located in an ego-defensive illusion of control, i.e. an exaggerated judgment of one's own grip on the circumstances (Perloff, 1983). Such an illusion leads to a reduction of anxiety and a feeling of having more control over one's existence. We have already seen this latter notion in Seligman's work. Maintaining feelings of control over a situation is adaptive.

According to Janoff-Bulman (1989), there are three basic assumptions that are tied to people's estimates of their own invulnerability: 1. the world is benevolent, 2. the world is meaningful and comprehensible, and 3, the person sees oneself as competent, decent and worthy. The victimization invalidates these fundamental beliefs. Victims, therefore, are faced with a dilemma: they must reconcile their prior assumptions (in part, illusions) that are no longer adequate and the negative experience that is too overwhelming to ignore. Therefore, they must revise and rebuild their basic assumptions.

The feeling of invulnerability is nullified by a traumatic event. Suddenly someone realizes that he is neither safe nor invulnerable, that life is uncertain and that he is not in control of his existence. He can be stricken by the same event again at any time.

This is one of the pre-eminent changes in self-image which Horowitz discusses in his theory. For the completion of the coping process, it is necessary to build up a feeling of security and safety – a more realistic awareness of invulnerability. An individual recovers faster when he sees himself and others as equally vulnerable in this unpleasant situation: 'it could have happened to anyone'. A perception of unique vulnerability – the feeling of being especially vulnerable, much more so than other people – is less conducive to the coping process. Perloff's study proved that this form of vulnerability was closely related to anxiety, depression and low self-esteem. Here we once more see the role of causal explanations. However, as opposed to findings mentioned above (Bulman & Wortman, 1977), it is not the attribution to internal or external factors that matters, but differences in the experience of being a victim. The awareness that other people run an equal risk proves to be a more comforting thought than the idea that the traumatic event is directed at the individual, as is the case with unique vulnerability.

People compare themselves to others when regaining control and when re-establishing a certain invulnerability. This social comparison is functional. Taylor's work showed that cancer patients increased their self-esteem, by comparing themselves to people who were worse off. Almost purposefully, they searched for someone who was in a worse state.

Moreover, the social comparison illustrates that the coping process does not take place in a social void. Such comparison provides the individual with information and advice and stimulates him not to solely blame himself for the causes. This counters internal attributions that are too strong. The mere awareness of the worries of others in a similar situation, can eliminate a notion of unique vulnerability.

Control and meaning

We have discussed various phenomena that pertain to regaining a grip on the situation after a traumatic event. What exactly is the difference between the search for control and the search for meaning? In the reformulation of the theory of learned helplessness, for instance, Seligman introduced the search for meaning, albeit a limited search for meaning, together with causal explanations for the event. Both elements of control and meaning are present in the awareness of vulnerability.

When control is defined as an active command of the circumstances, there is a difference between the themes; however, the concept has acquired an ever broader meaning, so that it overlaps with the attribution of meaning. When one is able to interpret one's life events in a satisfactory manner afterwards, then one still has a grip on the events. This was clearly shown in Taylor et al.'s study (1984), in which cognitive control emerged as being very important.

We can therefore state that attributing meaning to a situation is a cognitive, internal form of control. This view can already be found in the psychoanalytical literature on traumata. It was indicated several times that Freud (1920) explained the reexperiencing and the traumatic dreams of war veterans as attempts at regaining control.

This opinion about control also emerges in other publications. Rothbaum et al. (1982) define control as the influencing of both external realities and the personal consequences of these realities. They make an important distinction between primary control – the shaping of physical, social or behavioral reality – on the one hand, and secondary control – the adaptation to the existing reality – on the other. In the latter case, someone leaves the situation unchanged, but exerts control over the psychological effects. A good example of secondary control is tying oneself, in one's mind, to other people and groups in order to share in the control they have. The identification with powerful people gives one a feeling of control.

Although Rothbaum et al. (1982) distinguish a number of types of secondary control, interpretational control is the most important type for our argument. This concerns the attempts to comprehend or construct the existing circumstances in such a way that it offers a sense of meaning. This interpretation, this process of attributing meaning, creates a feeling of control and thus promotes well-being.

It is striking that these two main forms of control can be found in Lazarus' dichotomy of the concept of coping (1981). He distinguished:

1) problem-oriented coping: changing the disturbed relationship between the individual and the environment (direct action, for example), and
2) emotion-regulating coping: the regulation of the disturbed relationship, without tackling the causes themselves (mostly intra-psychic processes such as denial).

It can be concluded that both concepts of control and coping with stress are closely interconnected.

We do not wish to claim that primary forms of control after a traumatic event are insignificant. Lazarus (1981) in particular has shown that problem-oriented and emotion-regulating stress coping are present in every situation (Kleber, 1982). After a violent crime, for instance, someone can move to a safer environment; his feeling of vulnerability then decreases. Altered circumstances as well as new behavior create another situation, thus altering its 'appraisal' as well. It is clear that such activities can promote the coping process. The same may occur after a disaster or a serious loss (chapter 7).

In spite of this, it should be clear from this discussion that secondary control, specifically the interpretative form, is important in coping with traumatic stress. The situation cannot be changed. Exercising control, therefore, is often a process of meaning attribution. The search for meaning and control are two closely interwoven processes so that one may wonder whether or not they are identical. However, neither of them can be disregarded (for the time being) as it is not clear how the two can be reduced to one. Control is the most comprehensive term, but further conceptual analysis is needed to conclude whether or not the content of the process of meaning attribution as a whole can be described as a form of secondary control. Both issues can best be considered strongly overlapping, yet not quite identical motives in the coping process.

8.6 Conclusions

In this chapter, the psychological consequences of traumatic events were the subject of a general discussion. The first part of the chapter showed that the psychological phenomena after a traumatic event can be considered as reactions to what one has lived through, on the one hand, and as attempts to take away the sorrow and to build up one's existence again, on the other. These two perspectives are two sides of the same coin. A human being reacts to a traumatic situation and, at the same time, he copes with it. For instance, the avoidance of the location of the event is both a response to the shock and an attempt to cope with the anxiety.

First, a model was outlined to reflect the origin of the reactions and the factors that influence them in terms of the stress perspective. This is strongly geared to the analysis of the various reactions to burdensome and threatening circumstances. The central starting point was that people do not react in the same manner to such situations, not even when they are traumatic. All kinds of mediating and conditioning factors play an important role in the determination of the reactions. These factors will be dealt with in the following chapter.

The model portrayed the relevant determinants adequately but fell short in analyzing the process of coping with a shock. Initially, the various influences were stressed, but the emphasis then shifted to the processes in and through which a person adapts himself to a new situation.

This two-directional approach is typical for a transactional perspective. Transactional means that the analysis relates to the interplay between the individual and the environment; in this case a relationship in which the demands of the environment exceed the individual skills. Person and environment are viewed as systems that interact, and not as two separate elements. This perspective forces researchers to avoid presupposing linear causality. The decision to choose variable A as an independent and variable B as a dependent variable, depends on where the researcher intervenes in the processes. It is equally possible – at least from different perspectives – to begin with variable B as a determinant of changes in variable A. The processes studied are cyclical.

The process of meaning attribution is of central importance in transactional perspective. It is an important link in the chain of event and effect. The concept of 'appraisal' formulated by Lazarus (1981), clarifies the process of meaning attribution. The discussion of this concept formed a bridge between the first and the second part of this chapter, because 'appraisal' also refers to the individual taxation of the opportunities for dealing with the situation.

In this chapter we used an approach based on psychodynamic theory, stress theories and cognitive psychology. This does not imply that other perspectives are not relevant. After all, we are dealing with complex phenomena that can be viewed from different angles. Learning theory models (chapter 11) and psychobiological models (Van der Kolk, 1988) have been suggested. We maintain the opinion, however, that the analysis presented here provides the most adequate explanation of the coping processes and the multiple determination of these processes.

We strongly emphasized the disruption of existing ideas and expectations, frequently implicit assumptions of a self-evident nature. In this light, the term schema was explained as an abstract structure of organized knowledge, which applies to behavior and experience. A similar approach is presented by McCann and Pearlman (1990). In their model, these authors hypothesized cognitive schemata that are affected by traumatic life experiences and that, in turn, influence the way the extreme events are interpreted and responded to.

The process of coping with a traumatic event can, on the one hand, be considered to be an attempt to find meaning in what has happened; among other things, this implies the search for an answer to the question 'why me'. On the other hand, the coping process can be considered an attempt to regain control over one's own existence. A way of curbing the chaos and of ordering the world is to interpret the circumstances in such a way that a grip on the situation is regained.

In one way or another, the event and its implications must be integrated into schemata. At first, the shock can barely be accepted (the bewilderment phase); the interpretation of what has happened is halted. Subsequently, the event is alternately denied and reexperienced. This need not always be negative, because the exploration of the implications of the event finds expression through this alternation. Finally, integration takes place.

CHAPTER 9

DETERMINANTS OF TRAUMA AND COPING

A perspective that merely focuses upon intra-psychic processes is not sufficient when one seeks to understand the psychological phenomena resulting from extreme stress. The coping process does not take place in a vacuum but is influenced by the demands and restrictions of the situation of the event, the individual and the social context. These factors are responsible for the differences in the reactions to traumatic experiences.

In this chapter, we present an overview of those determinants that have been associated, in research or theory formation, with the process of coping with extreme stress events. It is our aim to structure this still largely unexplored area. We will concern ourselves with:

- the characteristics and circumstances of the event;
- the characteristics of the person involved;
- the characteristics of the social and cultural conditions.

9.1 Characteristics and circumstances of the event

A person is involved in an event that takes place in a specific manner, under particular circumstances, and at a given point in time. Various characteristics of the situation turn the event into an experience of powerlessness, disruption and discomfort, thereby affecting the consequences and the coping process.

The study of traumatic experiences is primarily focused on the consequences of an environmental element. However, situational determinants have been neglected in research. There are few results that can be presented.

An example of a disaster, where environmental issues play an evident role, is the Buffalo Creek flood discussed earlier (chapter 4). This disaster unexpectedly hit a community at night. The number of dead was high, the material damages immense. As access to the area was problematic, aid was slow to arrive. Long after the catastrophe, evidence of the disaster was still visible. All these circumstances served to make this disaster even more of an experience of powerlessness, disruption and discomfort.

Bowlby (1980), in his analysis of determinants of the grief process, mentioned the following situational factors:

- Did the death involve mutilation?
- How was the person informed of the other's death?
- Was there prolonged illness before death?
- Did anyone seem responsible for the death?
- How was the relationship prior to the death?

In chapter 7 we pointed out a determinant that is specific to the grief process, namely the nature and quality of the relationship between the deceased and the person left behind. The closer the relationship between the deceased and the individual, the more difficult it will be to cope. Strictly speaking, characteristics of a relationship between two people are perhaps not to be conceived as situational variables, but we wish to stress here that a loss is never merely a loss, but a specific loss in a given situation.

Characteristics of the event and its context, i.e. the severity of the stressor, are highly decisive factors in the process of coping with traumatic stress. Surprisingly, the assessment of the intensity of the stressor has hardly received substantial attention in traumatic stress research. There is considerable consistency across studies in showing significant correlations between combat exposure on the one hand and postmilitary psychological adjustment and posttraumatic stress disorders on the other hand (Foy, Carroll & Donahoe, 1987; Kulka et al., 1990). The frequency and intensity of various stress reactions after a severe explosion in an industrial factory were linked to the intensity of the exposure to the impact of the disaster (Weisaeth, 1989b). But in an Australian study of a bush fire disaster (McFarlane, 1988) intensity of exposure was not found to be a predictor of posttraumatic stress disorder. These contradicting findings are typical for the research on the etiology of traumatic stress reactions.

We would like to emphasize that a careful analysis of the characteristics of the situation of the traumatic event is necessary in order to understand the meaning of the event for the individual. Various aspects of the situation should be taken into account. We will focus here on two of these aspects.

The anticipation of an event

All traumatic experiences described in this book are unexpected. The onset of the event is virtually always sudden and is not, or barely, anticipated by the affected person. Nevertheless, differences exist in the degree of anticipation. In some cases, a certain degree of anticipation of an impending disaster is possible: circumstantial features make it possible to foresee the occurrence of the event.

Numerous stress studies have shown the impact of anticipation upon human behavior (Janis, 1971; Lazarus, 1966). Not only can it lead to insecurity, avoidance and fear, but also to various behavioral and emotional patterns that prepare the individual for an impending disaster. When a shock is anticipated, the individual can prepare himself for the deleterious event, not only materially but also emotionally. Experimental stress research shows that people who did not see or barely saw the event coming, experience more symptoms afterwards and are less able to deal with the changed circumstances (Lazarus, 1966).

The significance of anticipation was pointed out several times in previous chapters. Expected deaths, though extremely threatening, create fewer problems, also in the long run, for those who remain behind (Parkes, 1972). It is not surprising, therefore, that pathological grief reactions appear less frequently after an expected death (Stroebe et al., 1981). A finding from the Harvard study by Glick et al. (1974) was particularly striking: two to three years after the loss, none of those who had lost their partner suddenly had remarried, as opposed to 65% of those who had been prepared for the partner's death.

Anticipation does not only play a significant role in grief but in other traumatic events as well (Wolfenstein, 1957). The Dutch physician Cohen (1953) gives a convincing example in his study on concentration camps. Those who were aware that something ominous was awaiting them in the camps experienced less bewilderment and intense emotions than those who were not prepared for the horrors when entering the camps. It was as if the first group had armored themselves.

In this context, we would like to refer to Janis' work (1971). He studied patients' behavior before and after concerning surgery and divided the patients into three groups, based on the degree of fear they exhibited. Those who were moderately afraid of the operation beforehand suffered less pain and discomfort and exhibited less anger and irritation than those who had either experienced much or very little anxiety before the operation. Stimulated by a certain amount of anxiety, the former patients had been able to dwell on future events that awaited them after the operation and on what they could and could not do in that situation. The significance of what they were about to go through dawned on them. Those who had been very afraid avoided the operation in all its facets, while those who had experienced no fear before the operation did not dwell on the surgery either (because they, according to Janis, attempted to deny it). Neither group had been able to prepare themselves.

When people, driven by a moderate degree of anxiety, occupy themselves in thought and behavior with what is about to happen to them, they gain information that enables them to cope more adequately with the unpleasant situation. Relevant information is therefore the primary component in preparing for an impending disaster (Baum & Singer, 1982). Fear is one of the motives behind the seeking of information.

Anticipatory information alters the interpretation of the threat and draws attention to the various ways of how to deal with the detrimental effects of stress events. Information affects primary and secondary appraisal (chapter 8). Both of these processes are present in the anticipation of serious situations. The person can more adequately define the new situation and is able to consider means of coping with stress.

The research about the importance of the anticipation of stressful events has led to the development of the technique of stress inoculation, which aims at preparing people for upcoming stress. It will be discussed in chapter 10.

A person's own active behavior in the traumatic situation

In closing this section we wish to draw attention to a variable which, although mentioned very infrequently in the literature, is an important aspect of coping with trauma, and which once again stresses the relevance of the situation in which the event takes place. By definition, an extreme event is to be conceived as a situation in which someone is extremely powerless. Sometimes an opportunity for personal initiative does exist, however. It seems beneficial to the coping process when someone is active during the event.

Possible explanations for this phenomenon are either the awareness of control over the situation, or the release of motor (hyper)activity. Considering the lack of research on this topic, however, it is not clear whether it is caused by a cognitive or a physiological factor. In line with Rachman, we are inclined to consider the effect of a person's own activity primarily as an indication of control. Those who had been given a socially responsible task during the German air attacks on English cities, demonstrated a remarkable increase in courage. Rachman (1978) wondered whether or not the opposite of learned helplessness (chapter 8), occurred here. Forced by the situation, people became active, which in contrast can be labelled as 'required helpfulness'.

This factor, however, is not solely a characteristic of the event. Whether a person behaves in an active manner during a traumatic situation depends both on the circumstances (to what extent do they permit active behavior?) and on the person (to what extent is he willing or capable of activity?). Both the available opportunities offered by the situation and the personal skills play a role in the ability to be active.

Concluding remarks

As most researchers focus on patients, and thus on people with disturbances, characteristics of the event and its specific context are strongly neglected. This may have as a consequence that the reactions are 'psychologized'. Factors that relate to the responsibility of the individual are overemphasized, and the external conditions which also influenced the disturbances are disregarded (Smith, 1980).

In future research more attention should be given to a thorough analysis of the situation. What did the traumatic event entail? Under which circumstances did it take place? Attention should be paid to a systematical assessment of the diversity of specific characteristics of a situation. An event derives its traumatic character from the combination of these characteristics. The principle of multiple determination is an important explanatory factor.

9.2 Personal characteristics

Each person reacts differently to a traumatic situation. One individual manages to get over the event without difficulty, another experiences complaints for some time, and yet another suffers from psychological problems that make his life a hell for many years ahead. Personal characteristics and idiosyncrasies contribute to these differences.

The diversity of factors, as well as the incidental attention paid to the unclearly defined issues, have resulted in inconclusive and sketchy research. Although many authors point to the importance of personal variables, their relations to the consequences of traumatic events have not been studied explicitly or systematically. It is therefore useful to present a 'common sense' description of the skills or capacities a person needs in order to cope with a traumatic event and its implications.

After a traumatic event, a certain flexibility seems necessary as the person has to change some of his expectations, opinions and behavioral patterns, and replace them with new ones (chapter 8). Flexibility is also important in relation to the stress response syndrome formulated in chapter 2. Denial of the event and its negative implications is often adequate, yet, at other times, the reality of the situation must be confronted. Being able to alternate between these states is conducive to the coping process. Neither merely denial nor merely reexperiencing the event are beneficial to this end (Lazarus, 1983).

Essential is a certain personal competence in coping with difficult circumstances (Boeke, 1984), in other words an awareness and confidence that one can influence the situation. The requirements for coping with extreme stress, defined in terms of a 'common sense' psychology as flexibility and resilience, make it possible to assign meaning to a traumatic experience and to regain control over one's own life.

There is little point in dealing with all personal variables that could possibly relate to our subject. We limit ourselves only to those variables that have been studied explicitly in empirical research on traumatic stress. These personal factors are classified as follows:

1. Biographical aspects, such as age, sex, and socioeconomic class;
2. Recent life history variables, such as recently experienced stressful life events and previous confrontations with situations similar to the traumatic event;
3. Personality traits, which are subdivided into two categories: a) dispositions originating in specific experiences during childhood and, b) general personality characteristics.

Biographical characteristics

Important biographical data are age, level of education, socioeconomic class, marital status and sex. Research within the social sciences has shown that these variables are related to the occurrence of psychological disorders and illnesses.

As age, gender, education and socioeconomic class have been studied explicitly in relation to traumatic events, our attention will focus on these biographical factors.

Age

Each stage of life is characterized by its typical experiences, activities and tensions. An individual will therefore react differently during each stage, while it is evident that certain traumatic events occur more often during some stages of life.

One might expect resilience to decrease with age, as well as the ability, mentioned in the introduction of this section, to adapt in a flexible way to changed situations. On the other hand, one might just as well assume that 'wisdom' and thus also flexibility and resilience increase during the course of a lifetime.

The younger a woman – research usually concerns women – at the time she is widowed, the more intense her grief will be and the more her health will be affected. Young widows visit their general practitioner more often and use more medication. They indicate more emotional problems during the first 18 months after the loss (Parkes, 1972). In women and men under 40 years of age, pathological grief is relatively more frequent.

The death of a young person happens to be more unexpected than the death of an old person. It does not fit into existing expectations and ideas, in other words, into the individual schemata. Finding a framework of meaning (chapter 8) is more problematic. Moreover, showing grief is more acceptable and regulated among older people (Parkes & Weiss, 1983).

This inverse relationship between age and pathological grief has not always been confirmed (Bowlby, 1980). In the process of coping with the loss of one's own child, the influence of age is virtually insignificant. Moreover, we are dealing with a gradual difference; pathological reactions are certainly not restricted to young people (Parkes, 1972). Loneliness and anxiety are found to be greater problems for older widows (Sanders, 1988).

Although hardly any effect of age has been mentioned in research on violent crimes, the Dutch study on ex-hostages (Bastiaans et al., 1979) showed the same inverse relation. Older people suffered less negative effects from the hijackings. The explanation for this was that they felt supported by the idea of having had a useful life and could thus accept the situation more easily. Moreover, they could better predict positive outcomes that might eventually arise.

With regard to disasters, due to the nature of this specific stressor, the few existing research findings point in the opposite direction. Older people tend to receive a warning for a disaster later and often not at all, as they do not frequently leave the house and are less often part of informal groups. Afterwards they have more injuries, confusion among them is greater, and they feel deprived. It is more difficult for them to start all over again after material devastations, and they consider their losses to be greater (Catanese, 1978; Friedsam, 1962). This effect of age is mediated by factors such as the possibility of anticipation, the availability of social contacts and the nature of the damage.

Leopold and Dillon (1963) studied the survivors of a ship explosion a few years after the accident. The psychological complaints were clearly higher in older people than in young ones. The soldiers who broke down due to combat exhaustion in World War II were also somewhat older (Brill, 1967). This relationship has not been confirmed in the case of the American veterans of the Vietnam war, however.

Finally, an often mentioned assumption is that coping with captivity in a concentration camp becomes problematic when the survivor reaches a higher age (Bastiaans, 1974; for an opposite finding Antonovsky et al., 1971). Children leave home, work becomes less important, people are left more to their own devices and memories tend to emerge. This explanation does not point so much to age, however, as to the effect of age related role transitions, such as retirement.

Gender

It is not clear whether gender differences occur in reactions to extreme events. During the first year after a flood in Bristol, England, illness and death figures had risen more for men than for women (Bennett, 1970), but in a similar Australian study these differences were not confirmed (Abrahams et al., 1976). In a Swedish study (Leymann, 1988) female employees reported more stress symptoms after bank robberies, but in a general population survey of posttraumatic stress disorder no association with gender was discovered (Helzer, Robins & McEvoy, 1987).

Women are highly over-represented in studies on patients, but this does not necessarily mean they experience more problems. Women turn to mental health institutions more readily, admit to emotional problems at an earlier stage and accept help more easily (Martin, McKean & Veltkamp, 1986).

Gender differences have been studied in relation to pathological grief. After a serious loss, women experience more anxiety, use more pills and are admitted to hospitals sooner (Bowlby, 1980). The number of female patients in psychiatric hospitals whose problems had started after the death of the life partner, was seven times higher than was to be expected on the basis of the entire British population; for men this number was four times higher (Parkes, 1972).

An analysis of findings on gender differences in bereavement research (Stroebe & Stroebe, 1983) showed, however, that widowers suffer more from the loss of the spouse, when this is controlled for confounding variables. This is also supported by the finding that although widowers (in absolute terms) have slightly fewer complaints than widows during the first year, during the second, third and fourth year, they experience more symptoms. The coping process is more sluggish for them (Bowlby, 1980).

An explanation for the above should be sought in social relationships. Outside of work, men seem to establish fewer social contacts. They often leave the initiative to their wives, who have extensive informal networks of relationships. As a consequence, the loss of support is more profound for men in traditional marriages (Stroebe & Stroebe, 1983).

Education and social class

These two interrelated characteristics have incidentally been associated with traumatic events. In the Dutch study on ex-hostages (Bastiaans et al., 1979), a correlation was found between level of education and various negative effects of the hijackings: the higher the education, the fewer problems were experienced. People with a high level of education were said to have more skills and more self-confidence and could therefore control the situation more adequately. However, in one of the few other studies in which this variable was studied explicitly, this relationship was not confirmed (Leopold & Dillon, 1963).

In general, people from the lower social classes have more psychological problems, according to many studies on social factors and health, even though there has been an extensive discussion about the direction of causality in this context (Dohrenwend & Dohrenwend, 1981). It bears great relevance for the current discussion that the relationship between class and psychological disturbances can be linked (Kohn, 1972) to concepts such as lack of control over one's own existence, learned helplessness and lack of flexibility. One has less opportunity in the lower social classes to acquire important skills for coping with stress.

This effect of social class is expressed in the following research findings. Women with financial problems did not recover as quickly after being raped as women without those problems. Their lack of resources prevented them to bring about changes in their personal life that they considered to be important, such as moving to another residence (Burgess & Holmstrom, 1978). Poor adjustment following bereavement was found to be a consequence of an insecure economic position of the widowed person (Sanders, 1988). Symptoms of combat exhaustion in World War II and the Vietnam War occurred more in soldiers from the lower social classes (Brill & Beebe, 1955; Rachman, 1978). There were more of them at the front lines, however, and this resulted in a greater risk of combat exhaustion. Here we once again see that the influence of a demographic variable is mediated by other factors.

Most relationships between biographical variables and coping with extreme stress are not very strong. Our conclusion is that even though biographical variables are associated with the consequences of extreme experiences, they should constantly be linked to intermediate variables. In short, their influence depends on various specific and mediating factors: the nature of the event, related expectations and perceptions, behavioral obligations and social circumstances. In other words, they should be translated into specific factors which are closer to individual behavior (Jessor & Jessor, 1973).

Recent life history of the person

Are people less able to cope with traumatic experiences when they have lived through various stressful life events during the previous period? An answer to this question is provided by the 'stressful life events' approach (Holmes & Rahe, 1967). The central notion of this approach is that various recent events in one's personal life – marriage, moving house, loss of a job, exams, and so on – require adjustment from the person involved, and this increases vulnerability to illnesses and other problems.

It has become evident from many studies that, indeed, the number of 'stressful life events' during a certain period – the most predictive strength was yielded by a period of six months – is related to the occurrence of all types of illnesses and problems (Thoits, 1982).

On the basis of this approach, we may assume that coping with traumatic experiences will be negatively influenced by the number of such events preceding the actual shock. This assumption has been substantiated in studies on the consequences of traffic accidents (Foeckler et al., 1978) and of rape (Ruch, Chandler & Harter, 1980). No such relationship was found, however, in another study on recovery after rape (Burgess & Holmstrom, 1978). In his study on London widows, Parkes (1972) concluded that life events increased the chances of disorders in the coping process in the case of a serious loss, but it did not prove to be a very important factor.

The possibility of a positive effect of previous stress has been suggested, among others, by a finding that the experiences prior to the event do not relate in a linear but in an curvilinear manner to the severity of the consequences. A study on raped women (Ruch et al., 1980) showed that women with an average amount of recent life changes experienced fewer problems than women with many or few life events in their personal life. Experiencing a manageable number of stressful situations is thus considered to have a beneficial effect on the process of coping with a severe experience.

The conclusion that recent life events are not an important factor in coping with a severe loss, is confirmed by reviews of studies on stressful life events and health (Rabkin & Struening, 1976). These events should be seen as variables that only influence health when coupled with other factors. The role of recent life events becomes evident in the work of the British social psychiatrist Brown (1979), even though his studies do not directly deal with traumatic stress, but with depression. It is the interplay of three different factors – recent stressful event, lack of support and loss as a young child – that increases the probability that depression will occur.

Childhood experiences, personality traits and traumatic stress

Is a human being less capable of coping with a traumatic event when he has had a stressful childhood? The answer seems to be clear. The assumption that a childhood trauma is hiding behind a recent trauma is common. Traumatic experiences in childhood are supposed to remain a part of the individual personality. They result in defenses that shield against anxiety and aggression, but at the same time they are detrimental to well-being under and after severe stress. It is here that we recognize the incompetence and rigidity mentioned earlier.

Certain childhood experiences create a predisposition towards disturbances in coping with traumatic experiences in adulthood. Soldiers who broke down during wartime more often came from homes with psychological difficulties (Brende, 1987; Brill, 1967) and from 'broken homes' (Kalman, 1977). The feeling of basic trust played an important role in whether or not a breakdown took place during combat. Soldiers who still suffered from all types of problems many years after the war, had often lost one or both parents before their thirteenth year (Archibald & Tuddenham, 1965; Nace et al., 1977).

A connection with depression can be made here. Many studies have indicated a relationship between the loss of a parent in childhood and depression at a later age (Anisman & Zacharko, 1982; Goldney, 1981). Brown's work (1979, 1981) has already been mentioned. His studies show that women who have lost one parent during childhood, were more likely to suffer from depression during adulthood. This was only expressed when they had experienced a traumatic event – namely the loss of the husband –

and when, in addition, they did not feel supported by a trusted person in their environment.

Although depression is not the same as disturbances in coping, these findings do yield insight into the relationship between childhood experiences and later consequences. According to Wolfenstein (1957), adults can still carry anxiety-charged childhood conflicts with them, such as the fear of being abandoned and ambivalence concerning the parents. A shock may once again evoke these conflicts and anxieties. The person tends to interpret the current event in terms of his earlier experiences. Regaining a sense of control over one's own life will be difficult when someone has, in the past, learned that he can exercise no or very little influence over his life.

The relationship between childhood experiences and consequences of recent traumatic events should be viewed with caution. The relation has not always been confirmed empirically (Boman, 1986; Burgess & Holmstrom, 1978), not in the least because of methodological problems. Certain childhood experiences make an individual more vulnerable to problems and symptoms, but it depends on other factors whether or not that person will actually exhibit these symptoms after a traumatic situation.

It is therefore necessary to specify experiences from childhood and youth. Which events from the past obstruct coping with traumatic events in adulthood? Which predisposing personality traits are the result of earlier unpleasant experiences?

We will consider attempts at formulating these characteristics. Elaborating upon the concept of control mentioned earlier, we will discuss general traits from personality psychology, such as the expectancy of people about their influence on their lives. These traits have been related to stress. First, however, Bowlby's analysis of childhood experiences and predisposition will be discussed. Although it is limited to grief, it clarifies the specific role of childhood experiences.

Attachment and coping

Bowlby came to the conclusion that most people who experienced difficulty in coping with a recent loss were those who had always entered into certain types of affective relationships. He distinguished three types of relationships which these people tended to have. These three tendencies stemmed from childhood, especially from experiences within the parent-child relationship. For each disposition, Bowlby described a specific family background.

The first disposition concerns a tendency to enter into uncertain, ambivalent relationships. On the one hand there is a strong attachment to the partner, on the other hand, anger, irritation and fighting characterize the relationship. Parkes and Weiss' study (1983) showed that widows and widowers who could not cope well with the loss had more disagreements and more ambivalent feelings towards the partner during marriage. According to Bowlby, such an attitude stems from certain childhood experiences. The parents had experienced the child's need for care as a burden. They reacted with irritation and often threatened with abandonment: 'if you don't listen, I'll leave.' Often the child had different babysitters or divorced parents, so that the upbringing lacked continuity. The rejection of the child had, however, only been partial, so that the child kept hoping and striving for care and attention; it did make him insecure, however. Typically, such a child refused to be left alone. He was afraid of losing the people he was attached to, became angry quickly and felt dependent.

The second disposition refers to a tendency to compulsively take care of others. When such caring is taken away, an enormous emptiness remains. Self-reproach and exaggerated idealization of the deceased may result. Bowlby links two types of childhood experiences to the exaggerated tendency of taking care of others; on the one hand the confrontation with inadequate and inconsistent motherhood and, on the other, the pressure exerted to care for a sick parent.

A third disposition refers to the tendency to be overtly independent of affective ties, while that independence has an insecure base. In Maddison's study (1968), nine out of twenty grief cases displayed such a personality trait. This disposition stems from: 1. the loss of a parent during childhood, which left the child to its own devices, or, 2. an unsympathetic, critical attitude of the parent in terms of the child's need for care. In both cases, this led to a shielding off of feelings. The child received the impression that affection was considered to be childish and a sign of weakness and thus developed into a tough, competent person. A severe loss in adulthood may break down that very toughness and independence, so that a person collapses, often only after some time has passed.

Bowlby links pathological grief to three personality traits, which he then connects with childhood experiences. The attribution of meaning in terms of the parent-child interaction plays a central role in this analysis (see for a similar approach Van Uden, 1988). Bowlby's approach matches the theory of Kohut (1977), who described in detail how childhood emotional neglect leads to personality characteristics predisposing to disorders in coping with traumatic stress.

Although Bowlby does not explicitly formulate it as such, *ambivalence* is characteristic of all these dispositions. The ties with other people are characterized by an overt or covert alternation between dependence and clinging on the one hand, and independence and rejecting on the other. A mourner reacts excessively to the bereavement but at the same time attempts to avoid the loss as much as possible. There is no balance. Due to specific childhood experiences, flexibility and resilience are not present when handling tensions.

Bowlby's analysis, which is based on clinical impressions and empirical studies, is a clear step towards specification of the connection between childhood experiences and disturbances in adulthood, although its rather speculative nature should be emphasized.

Neurotic predisposition

Neurotic instability, defined and operationalized as a personality trait, is widely assumed to lead to disorders after a shock. There is some empirical support for this assumption. Parkes (1972), for instance, found a greater number of cases of pathological grief in widows and widowers who had previous psychiatric records. In the Dutch study of ex-hostages, the degree of neuroticism, measured by means of a standardized questionnaire, was related to the negative consequences of the hijacking (Bastiaans et al., 1979). Neuroticism and a past history of psychiatric problems were factors significantly associated with the development of posttraumatic stress disorder in firefighters exposed to a bush fire disaster (McFarlane, 1988).

In terms of the stress model outlined in chapter 8, this personality trait is a conditioning factor. When traumatic events occur, a high degree of neuroticism negatively reinforces the reactions. The chances of breaking down emotionally during combat activities in World War II, was seven to eight times higher for soldiers with neuroses than for those who did not suffer from emotional problems (Brill & Beebe, 1955). The consequences of accidents also proved to be related to the presence of already existing emotional problems (Andreasen, Norris & Hartford, 1971; Foeckler et al., 1978).

It is not justified, however, to conclude that a neurotic tendency is in itself a sufficient prerequisite (or even a necessary one) for the occurrence of problems after a traumatic experience. In chapter 3 it was shown that by no means all soldiers who had been diagnosed as psychiatrically unfit for service actually suffered from combat exhaustion during World War II, and in Swank's (1949) study, pre-existing emotional problems were insignificant as an explanation for combat exhaustion in comparison to a factor such as the number of dead and wounded in an unit. Also the study of ex-hostages showed that neuroticism contributed less to the prediction of negative consequences than the negative experiences during the hostage situation (Jaspers, 1980).

Terms such as 'neurotic predisposition' and 'pre-existing emotional problems are vague descriptions. How does the fact that someone has a neurotic character effect the process of coping with a traumatic event?

An explanation of the consequences of traumatic events in terms of neuroticism as a personality characteristic is almost circular in nature. Neurotic people exhibit by definition more neurotic symptoms, and posttraumatic stress disorder consists of neurotic symptoms. The same applies to characteristics such as trait anxiety and lack of self-esteem, which have been related to the negative outcome of stress situations (Burgess & Holmstrom, 1978; Defares, 1990). This circularity can be avoided when the variables are measured before the traumatic event, but such an assessment is hardly possible in traumatic stress research. The danger of a tautology is nonexistent when a trait explains behavior other than that from which the trait is derived. The latter is the case with the concept of 'locus of control'.

Locus of control

Although its operationalization has been heavily criticized, the concept of 'locus of control', originally developed by Rotter in 1961, is one of the most widely used and discussed concepts in modern psychology. It has also been analyzed in research on traumatic events.

'Locus of control' refers to the general feeling a person has about his ability to determine the course of his own existence. People with an internal locus of control believe that what happens in their life mainly depends upon themselves. People with an external locus of control assume that their life events are determined by God, fate or other external factors. The concept of 'locus of control' thus refers to the way in which causality is assigned.

In his social learning theory, Rotter (1966) utilizes this concept as a generalized expectancy. To avoid confusion, we point out that 'locus of control' is not identical to the concept of control in chapter 8, which refers to the attempt of regaining a grip on the situation by changing the environment or by thinking differently about the circumstances. Locus of control indicates the general feeling that people have about their ability to influence their own life and should be regarded as a personality trait, independent of situations.

Research findings about locus of control are summarized as follows. People with an internal 'locus of control' generally achieve more, are more successful at school and work, and try to shape their own lives more than people with an external 'locus of control' (Lefcourt, 1980). An external 'locus of control' is clearly related to anxiety and poor functioning. People with an external 'locus of control' more readily admit their failures and their problems; 'internals' forget failures more quickly, although they try harder to solve their problems. Their ability to cope with stress is more successful than that of more passive 'externals' (Phares, 1978).

The belief of control develops in childhood. It has been shown that support and warmth from the parents, as well as the fostering of independence are necessary preconditions.

Anderson (1976) studied locus of control and coping with stress in small entrepreneurs located in a disaster area that was hit by a severe flood. Eight months after the disaster, 'externals' felt that they had experienced more stress immediately after the shock. They handled the problems and tensions arising from the situation in a less task-oriented and more defensive manner. Business results in the stricken area were analyzed 2.5 years later in a follow-up study (Anderson, 1977). The stronger the feelings of internal control, the greater the achievements proved to be. Other studies (Sandler & Lakey, 1982) also showed that people with an internal 'locus of control' adapted to stress situations in a more task-oriented way. They attempted to acquire more information on the problems and used such information more effectively.

A study conducted by Sims and Baumann (1972) on control beliefs is quite remarkable. The number of dead due to tornados in the South is substantially higher than in the North of the U.S. This difference could not be explained by density of population, the number of tornados, the warning system or other 'objective' factors.

The researchers requested inhabitants from a southern state (Alabama) and from a northern state (Illinois) to fill in a sentence completion test. The sentences had been constructed in such a manner that it could be determined how someone dealt with the threat of disaster and what the direction of 'locus of control' was.

The inhabitants of Illinois proved to be more autonomous; they saw themselves as more responsible and relied more upon their own actions. Inhabitants of Alabama had less confidence in themselves, and considered themselves dependent on God, fate or luck. Furthermore, the inhabitants of the northern state reacted more rationally and more actively to a tornado and accepted the advantages of technology. The Southerners were more passive and disregarded technological aspects. Although there certainly were other factors which played a role, a sense of control was posed as an explanation for the clear difference in the number of victims.

Whereas the above mentioned studies concerned natural disasters, Frye and Stockton (1982) linked locus of control to the presence of difficulties of Vietnam veterans in coping with their war experiences. In addition to their family's support, and the nature of the combat in Vietnam, the sense of control proved to be the only factor that was related significantly to the occurrence of posttraumatic stress disorders. The more external the locus of control, the more disorders occurred. The following explanation was given by the researchers. In Vietnam, unpredictable and uncontrollable events were commonplace. Such events could have led to depression, but a history of internal control protected the individual against the effects of unpredictability and uncontrollability. It should be noted that control beliefs were measured in this study 10 years after return from Vietnam; it is possible that the disorders in coping with war experiences led to a sense of external control.

On the basis of the above mentioned findings, the generalized expectation that someone's life events are dependent on one's own actions, seems conducive to the process of coping with extreme stress. Those who believe that they are in control of their own destiny can cope better with trauma and work more actively – even if only in thought – to regain recovery (Lefcourt, 1980).

The latter statement seems self-evident, but the opposite may also be understandable. Attempting to exercise influence on something that cannot be influenced evokes distress. An internal belief of control is not always a condition for handling trouble more effectively (Lazarus, 1981). Someone who is terminally ill, or someone who has lost a loved one, must resign himself to the inevitable. Sometimes passivity is a more adequate response (Lefcourt, 1980). The nature of the situation is just as important as the person's expectations. Rigidly adhering to an internal locus of control may not be adequate.

As yet, however, the above mentioned phenomenon has not been studied very well. It is interesting to mention a study on burn patients (Andreasen et al., 1971). The researchers assumed that passive burn patients would have fewer problems with the isolation, passivity and dependence on the nursing staff. The opposite was true, however. Those who were active, successful and alert in daily life, proved to be more capable of coping. People with an internal control awareness could accept the conditions of illness more successfully.

Concluding remarks

An individual has developed a complex of meanings about the world, as well as expectations about his skills and resources in dealing with various situations. This framework is rooted in previous experiences, in his position in terms of development and environment, and in certain personality characteristics. These personal factors shape the process of coping with traumatic events. Building up new frames of meaning and regaining a grip on one's situation may either be hampered or facilitated.

However, the relationships between personal variables and coping with traumatic stress have not often been the subject of empirical studies; moreover, research is hampered by methodological and conceptual shortcomings. A serious shortcoming of research in this area concerns the scanty specification of personal factors studied with regard to traumatic experiences. Often the variables are formulated too generally. Another problem concerns the fact that personal variables can only be measured after the event, so that their real influence is difficult to determine. Changes in personality characteristics may be the consequence the shock itself.

Over the past 15 years, theory building within personality psychology has clearly moved into the direction of interactionism. Situation and person mutually influence one another (Endler & Magnusson, 1976). In terms of such a perspective, one might expect that personality traits may change through the traumatic experience. Parkes (1972) found no difference in various personality traits between widows and widowers on the one hand, and married men on the other. This would mean that such traits do not change due to a traumatic experience. Enduring personality changes, however, have been found among survivors of long-term traumatic situations, such as concentration camps (chapter 6).

Although research outcomes are not always convincing, it is evident that personal factors play an important role in the consequences and the process of coping with traumatic events. In future studies it would be useful to analyze the individual ways of coping with stress in a more detailed manner, preferably in longitudinal studies, in order to determine the individual differences and their determinants. It is also important to focus on the characteristics that are conducive to the coping process, and not only on the impeding factors. Too much attention is currently being focused upon, for instance, neurotic instability and earlier problematic experiences, while too little is focused upon the abilities and skills used for coping with stress.

9.3 Social dimensions

Social factors play a role in the process of coping with traumatic experiences. Psychiatric and psychological aspects, however, have predominated in traumatic stress research. Influences of the social environment have only been touched upon.

We look at the social dimensions of the consequences of serious life events from two angles. First, there is the influence of other people during and after the event. The coping process may be facilitated by support, information and encouragement or may be impeded by rejection, incorrect advice and discouragement.

In the second part of this section we discuss the impact of culture on coping with extreme stress, and focus in particular on the role of regulations and rituals that are characteristic of a society. A person is raised in a culture that has taught him to perceive, think and evaluate in a certain way. These cultural influences have become internalized by the person in the course of his socialization process and shape his responses to traumatic experiences.

These two types of influences from the environment are interrelated. Culture affects people's life in particular through small groups, and determines therefore the support that someone receives after a traumatic situation.

Social influences during and after traumatic stress

A human being is never alone when he is confronted with a traumatic experience. He forms part of a network of people, who can support, encourage and assist him. The importance of the social network during an extreme experience is shown by the scientific literature on soldiers during World War II. Feeling supported by a small, solid group proved to play an important role in perseverance during combat. Grinkel and Spiegel (1945) indicated that most of the American soldier's affection and energy was devoted to his unit. The group constituted his 'super ego' as it were.

One might wonder whether this is only typical of American soldiers. Riesman (1950) described the American culture as 'other directed', i.e. strongly geared to the opinion and behavior of other people. European cultures are said to be more 'inner directed', more oriented toward inner considerations and internalized values. Therefore, the primary group should be more central in American than in European culture. This distinction, however, has not been confirmed. The findings about American soldiers also applied to German soldiers. These were primarily motivated the small unit of fellow soldiers and a few officers of which they were a part (Rachman, 1978). Political and ethical considerations barely played a role in remaining morale within combat.

A study on American infantry men (Rachman, 1978) indicated that, for soldiers, the second most important reason to fight was the necessity to support and stand by their close comrades. The primary goal was obvious: the desire to end the war and to return home. Ideological, political and patriotic considerations were usually not important to soldiers, unless they had purposefully sought out combat, such as volunteers during the Spanish Civil War (Rachman, 1978).

The important role of the social network was proven most clearly by the behavior of soldiers while on their own. When one of them had become isolated from his group, this caused intense anxiety reactions. They lacked the support from others (Rachman, 1978). The worst thing during the Vietnam war, according to many, was to stand guard alone. In experimental studies on social affiliation in social psychology (Sherif & Sherif, 1969), it has been proven that the presence of others reduces anxiety.

These findings indicate that social support during a serious situation contributes to the appraisal of the circumstances as less powerless, disrupting or discomforting, therefore, resulting in fewer negative consequences. Social factors shape the appraisal of the situation (chapter 8).

The importance of social support, or the lack of it, after a traumatic experience, clearly emerges from a comparative study of two groups of Australian and American widows (Maddison & Walker, 1967). One group of women were still in bad shape 13 months after the loss of the husband while a group of comparable women had nearly recovered. The widows from the first group complained that the immediate environment had kept them from expressing anger and other emotions. They had not been able to talk much about the deceased. The widows of the other group mentioned that someone had shown understanding of their feelings. With that person – a friend, family member or general practitioner – they had been able to talk about the loss and had been able to discuss memories of the deceased. Their feelings and behavior were accepted; the pressure of having to be 'firm and strong' had been minimal. The immediate surroundings stimulated the expression of emotions, and, in that way, prevented pathological grief.

People who end up alone after the death of a loved one have more problems to contend with. In a study by Clayton (1975), 27% of the persons who lived on their own suffered from depression during the first year, compared to only 5% of those who did not live alone. Parkes (1972) reported comparable findings.

Similar conclusions may be drawn from some studies on raped women. The victimized women largely went to family members or friends for support. Fear and other symptoms lessened when a woman spoke about the situation with others (Sutherland & Scherl, 1970). The women who could not share their experiences with people from their immediate environment were usually the ones with serious complaints and symptoms (Evans, 1978). The same is true for the consequences of traffic accidents (Foeckler et al., 1978) and for disturbances after wartime experience (Barrett & Mizes, 1988; Foy et al., 1987).

Notes on social support

Social support is a concept that has received much attention over the past few years, especially in the area of stress research. It had, however, been studied earlier in such other areas as sociology (in studies on social integration and anomie), anthropology (in social network studies), developmental psychology (the role of the family) and social psychology (in reference group studies from the '40s and '50s).

Many studies on stress and health (Mueller, 1980; Sarason & Sarason, 1985) have found that social support correlated negatively with symptoms, illnesses and problems. The influence of support can be strong. A study by Berkman and Syme (1979) showed that the degree of support people experience is related to the duration of life; those who were deprived of social and community ties died sooner than those who had intensive ties.

The nature of the relation between support and health is not clear, however. Some studies indicate that social support neutralizes the negative effects of drastic events. It is a protection or buffer against stress (see Brown's previously mentioned studies). In this sense, social support is viewed as a conditioning variable, just like some personality characteristics. In other studies, a direct effect of social support on health and adaptation has been discovered. Aid from others prevents the occurrence of symptoms, regardless of whether there are many or few stress factors. The buffer effect and the direct effect prove to exist concurrently (Thoits, 1982; Winnubst, Marcelissen & Kleber, 1982).

An important issue concerns the nature of social support itself. What does it imply? Social support has different dimensions. There are various sources of support: the spouse, friends, colleagues, and the neighborhood. Aside from these sources in the individual's social network upon which he or she relies for aid, the support may take on different forms. The following types of support are distinguished in empirical research (Flannery, 1990; Hirsch, 1980; Moos & Mitchell, 1982; Thoits, 1982):

1. cognitive support, i.e. providing advice, information and explanations for phenomena;
2. social sanctioning, i.e. approval or disapproval of specific behaviors;
3. material help, such as financial support, but also other 'concrete' activities, such as helping around the house;
4. companionship or camaraderie, the support one experiences when doing things together with others, e.g. going to the cinema;
5 emotional support in a narrower sense, i.e. activities and remarks from others which make the person feel better when he is under pressure ('a pat on the shoulder').

Some forms of support are more important than others in a given situation. In a study on recently widowed young women, Hirsch (1980) asked about the degree of support experienced. The better the cognitive support, the less symptoms and the better their mood. The information and the advice enabled the young women to determine their new tasks and facilitated their coping with the complex changes. A higher self-esteem correlated with companionable experiences. Hirsch did not discover other correlations among the five forms of support and mental health variables.

In short, stress evokes different reactions and actions, which make the various forms of support more or less relevant. Therefore, support may have no effect or even an adverse effect. Stringent control from the immediate social environment may hinder a person in his functioning. In a study on women from rural areas in Scotland, Brown and colleagues (Brown, Davidson, Harris et al., 1977) found that the integration of women in the local community related to less depression, but also to more anxiety and insecurity. Parkes (1972) and Glick et al. (1974) mentioned that the social network did not exercise a positive influence on the process of coping with the loss of the partner when it consisted of small children who were not grown up as yet. It was difficult for the widow to take care of them by herself and she was hampered in starting a new life. The mere presence of others does not, necessarily, imply support.

The extent of social support does not only determine response to traumatic stress. It is also an outcome variable. Some traumatic events lead to changes in social contacts. A woman who was raped in the street is afraid to leave her home and will not easily look for support in the outside world. An event such as the death of the partner is, to a large extent, the loss of support. The intense emotions after a traumatic event may repel others (Lazarus, 1983), resulting in an increase of feelings of abandonment. Frequently family and friends are frightened and do not wish to hear about the disaster. In short, support and traumatic stress influence one another reciprocally.

For the future, social support should be analyzed in terms of its associated parts. A social network approach reflects more adequately the field of interpersonal contacts that exercise different positive and negative influences (Flannery, 1990; Hobfoll & Stephens, 1990). Social ties are of primary importance to human functioning but they do not always influence in a positive manner, as suggested in the concept of social support.

The impact of society

We have shown in chapter 3 that Vietnam veterans probably experienced more distress after their return to the United Stated than World War II veterans after 1945. These differences might have something to do with the nature of a jungle war in a hostile country such as Vietnam, but this factor appears to be insufficient to explain the many disturbances among Vietnam veterans, in particular because during the actual fighting in Vietnam few emotional problems arose.

A more satisfactory explanation for the many problems of Vietnam veterans is the social reception given to these soldiers in the United Stated, or, in other words, the political climate of American society in the seventies.

After World War II, soldiers came back as heroes. They had saved the country from the enemy. They were received everywhere with cheers and enthusiasm. When applying for jobs, veterans had priority over those who had not fought. The Vietnam veterans, however, had quite a different reception. At the end of the sixties and the beginning of the seventies, the U.S. were vehemently debating the sense of fighting the Vietnam war. Subsequently and with increasing frequency, public opinion turned against the American intervention. A massive anti-Vietnam movement existed. Many veterans felt estranged from the society to which they returned.

In addition, they came back by plane. After having served for one year, they left their combat unit and their comrades – sometimes in the middle of combat action – and flew directly home. Suddenly their network of intense social contacts – the so-called 'buddy culture' – was destroyed.

After the end of the Vietnam war, the estrangement increased. For most Americans, the Vietnam war was a painful and controversial matter, one that was better left unspoken. The Vietnam veterans were considered to be representatives of this abominable war. DeFazio (1975) mentions that nearly all veterans had the feeling that they were regarded as monsters by the rest of the population. There were few people with whom they could share their often horrible experiences.

Consequently, many veterans secluded themselves from the others. Bitterness and resentment frequently developed toward official authorities (Ewalt, 1981). Feelings of detachment and isolation as well as significant problems in the areas of intimacy and sociability have been reported among combat veterans (Peterson et al., 1991). As for the veterans themselves, the war and their involvement in it received a negative connotation, which resulted in low self-esteem.

The psychological atmosphere in society after the veterans returned was clearly a traumatic event. It may precisely be this climate that largely caused the problems. In his analysis of a disaster, Erikson (1976) spoke of two traumas: first, the occurrence of the disaster, and second, the destruction of community life and loss of social contacts. We see the same phenomenon with regard to Vietnam veterans. A similar view on the reentry of Holocaust survivors into society has been described by Keilson (1979).

We may state that the disturbances of the Vietnam veterans are determined at least as much by this socio-cultural trauma as by the trauma of combat. The enormous lack of recognition aggravated the problems for the veterans. Some research outcomes are interesting in this context. Seventy-five percent of World War II veterans felt they were well received upon their return; less than 50% of the Vietnam veterans had this impression (Ewalt, 1981). In a previously mentioned study, Frye and Stockton (1982) distinguished the group of maladjusted Vietnam veterans from the group without adjustment problems. The helpfulness and support of the family after the return from Vietnam was by far the most important factor that discriminated between the groups. All these findings point to the importance of the social climate in determining the intensity of psychological disturbances after extreme stress.

Culture and trauma

Man has been described as a 'nicht festgestelltes Tier' (Nietzsche); he is only partly determined by heredity. Culture supplies him with behavioral patterns, ways of thinking and feelings. Culture can be seen as an acquired 'lens', through which an individual perceives and understands the world that he inhabits, and learns how to live within it (Helman, 1990). Cultural elements form an intrinsic part of an individual's personality and behavior, although we are usually not aware of most of them. Because the impact of culture on coping with serious life events has primarily been studied with regard to grief, attention will be focused upon this field of research.

Although cultural customs can vary greatly, authors such as Averill (1979) and Bowlby (1980) are of the opinion that the feelings of bereavement and the coping process itself are the same in all cultures. Although customs vary, human reactions remain the same, as was shown in a study of bereavement in Japan (Yamamoto, Okonogi, Iwasaki & Yoshimura, 1969). As an example, Bowlby mentions the 'presence' of the deceased. Just as the lost spouse remains present in our culture through dreams, hallucinations or imaginary camaraderie, the deceased individual is viewed in other cultures as a dangerous demon or an ever-present forefather and spiritual support for the next-of-kin. Some authors (Averill, 1979) presume a biological basis for grief because monkeys and a few other mammals show reactions that strongly resemble human responses to the death of a loved one.

In an analysis of the ethnographic descriptions of grief reactions in 78 different societies, crying was the most frequently mentioned reaction of the relatives. In 76% of the societies, anger occurred, as did fear and automutilation. The authors (Rosenblatt, Walsh & Jackson, 1977) came to the conclusion that emotional reactions to bereavement are the same in all cultures.

Not everyone agrees with this universality. According to some (Volkart & Michael, 1977), the degree to which the loss of a spouse is felt and expressed varies greatly cross-culturally, among other things because the character of marriage varies greatly. Considering the death of a loved one as an irreparable loss is said to be a typically Western phenomenon. However, those who dispute the universality of grief reactions mainly refer to habits of mourning and not so much to the coping process.

Nevertheless, the loss of the spouse is experienced profoundly in Western culture. The individual's existence is dominated by only a few very intense and stable ties. The small family forms the basis of activity, attachment and affection for an individual, who is, therefore, very dependent upon these close ties. The loss of a spouse in our culture also implies the loss of income, status, security and even a part of one's own identity.

Grief is usually coped with alone or in a small group. In some cultures it is common to share feelings of grief with the entire community. In American and Western European society, it is more or less dictated to bear the loss in a controlled manner (Charmaz, 1980).

The restrained way of coping did not always exist in Western Europe (Ariès, 1974). In the Middle Ages, dying was a public ceremony, where everyone, including children, gathered around the death bed. The dying person expressed a number of requests, and played the most important role in the ritual. The French historian Ariès (1974) suggests that the subsequent mourning process was not a lengthy one. Death was always near and therefore self-evident.

After certain customs, such as wearing mourning clothes and visiting the closest relatives, had come into existence during the Middle Ages, it was not until the 19th Century that excessive emotional displays (e.g. fainting and crying) came into being. During the Romantic Movement, an almost morbid interest in death grew. Only then did the regular visiting of graveyards come about, so that the deceased was kept increasingly alive.

In our history, Ariès sees an ever increasing problem in accepting another's death. Nowadays, dying takes place in a hospital with the children usually no longer present. The physician, actually an emotionally uninvolved bystander, has become the most important person. Mourning clothes disappear. Funeral services are conducted in an increasingly detached manner. This does not indicate, however, that people are more indifferent with regard to death. In the past, the display of grief was quickly replaced by resignation. Today, people cope with bereavement alone. According to Ariès, it is striking that widows and widowers remarry much less these days than before, despite the minimal display of grief.

Ariès was strongly inspired by the famous study of Gorer (1965), in which the consequences of the death of an important relative were examined in 80 British men and women. Gorer fiercely criticized the lack of adequate grief rituals in British society. The cultural belief systems and practices that regulate grief and behavior have disappeared. The loss of a loved one has become a much more dramatic event because of the poor social support and the inability to express sadness. In particular in Great Britain, a wake is no longer held and the deceased is usually no longer placed on a bier. It is ironic, that many studies concerning grief and the problems of death have been conducted by British researchers.

Gorer blames the decline in mourning behaviors and rituals for the many cases of pathological grief. Although this premise is extreme, it does indicate an important connection. It is remarkable in this context that Parkes and Weiss (1983) in their study on grief found that widows who showed no grief, wore no mourning clothes and did not visit their husband's grave suffered from more disturbances than other women three months after the loss. A direct connection between religion and the occurrence of healthy grief has not been proven, however (Van Uden, 1988).

The influence of culture on posttraumatic stress is not known, because it has hardly been investigated yet. Most medical anthropologists (Helman, 1990) would agree with the general view that there are certain universals in abnormal behavior and that western categories of major disorders, in particular psychotic disorders, are found throughout the whole world. However, there is a wide variation in their form and distribution.

Each culture provides its members with ways of shaping their suffering into a recognizable illness entity and of explaining its cause and treatment. Culture determines the language in which psychological distress is communicated to other people. Depressed patients in non-western cultures often complain of a variety of diffuse physical symptoms, such as vague pains, headaches, dizziness, and general malaise. Somatization represents a dominant way of coping with unpleasant experiences in many societies. Individuals interpret and articulate personal and interpersonal problems, and experience and respond to them, through the medium of the body (Kleinman, 1986). High rates of somatization have been found in many studies of depressive patients in various African, Asian and Mediterranean countries. The effects of distress are expressed in a non-psychological language, while psychologization (the use of psychological terms, such as loss of meaning and emotional complaints, to describe subjective mental states) is more common in western societies.

Posttraumatic stress disorder is a concept that has been developed exclusively in the western world. Just like depression it contains many elements of psychological distress. It is therefore uncertain whether complaints and problems of non-western patients after extreme stress are covered by this concept. There is a danger of failure; appropriate diagnosis may be missed and treatment may be withheld because of the façade of somatic symptoms.

Cultural dimensions have been the subject of some research, not only about grief, but also concerning combat exhaustion. Williams (1951) was involved as a field psychiatrist in treating British and Indian soldiers who had suffered breakdowns in the Birma war against Japan. Over a period of 18 months, he examined more than 600 soldiers suffering from combat exhaustion, and was able to investigate the differences and similarities in complaints between people from different cultures.

In Williams' research findings, it is clearly shown that combat exhaustion was a general phenomenon, but that differences existed in the ways in which soldiers from different cultures expressed their breakdown. The British soldiers acknowledged far more symptoms of anxiety. Loss of face because of fear was an extremely shameful experience for the Indian patients. Characteristic for the latter soldiers were the frequent occurrence of hysterical symptoms of a rather extreme form (expressed in aggression, paralyses, and fits of anger) as well as the occurrence of brief psychotic episodes. Finally, Indian soldiers reported more physical health problems.

Unfortunately, only a few studies have been conducted on combat stress reactions in for instance German, Russian or French World War II soldiers. It shows that social and cultural factors have not received much attention in the field of traumatic stress.

Psychological disturbances after extreme stress have recently been studied in special populations (refugees and migrating minorities) in the western world. Kinzie and Boehnlein (1989) have described disorders and issues of treatment with regard to refugees from Cambodia who suffered massive trauma during the Pol Pot regime. Posttraumatic stress disorder was found accompanied by other psychiatric disorders, such as substance abuse, depression and psychosis.

Concluding remarks

Social support is a simple and appealing concept: supported by others, man can overcome misery. However, we are dealing with a complex phenomenon. Support may be given by different people or groups, varying from the spouse to the whole of society. Support may come in different ways, from encouraging remarks to financial help.

Support by other people influences the interpretation and handling of the situation. It influences primary and secondary appraisal (Lazarus, 1981). The event is appraised as less traumatic, and information is provided on ways of coping with stress and one's own skills. Support may foster self-esteem and counter the negative consequences of the traumatic event.

The positive effect of social support, however, is not always present. Negative effects may occur depending on its context. Moreover, support is not only an independent variable. A traumatic event, like the death of the spouse, implies the loss of an important source of support, with possible consequences of loneliness, lack of assistance in activities, and loss of material support. A person looks for help from others after a drastic experience, and this search for support may be considered as a form of coping (Defares, Brandjes, Nass & Van der Ploeg, 1985). A dynamic process is taking place between social support, traumatic stress responses and coping.
In future research, it would be useful to give attention to the following:

1. the social network of the individual, i.e. the ties and interactions with others;
2. the way in which a person experiences these ties and the extent to which he feels supported by them.

Although examined less than social support, cultural views and practices are also important. They provide the channels, as it were, in which the emotions of the coping process are expressed in socially acceptable ways. In terms of what was contended in chapter 8, they offer schemata to the individual. Rituals give a general meaning to traumatic events, i.e. they provide a cognitive framework. Smith (1980) indicated that society shared the responsibility for the killing of the individual soldier ('I did it for my country') in rituals that took place directly after the end of World War II (the parades, the receptions, the heroics, etc.). Cultural practices and rules also shape the interpretation and handling of the traumatic event and thereby the coping process and the consequences.

In examining these cultural variables, we have discussed the question to what extent the coping process is culturally determined. Although little research into this matter was conducted, the impression exists that the process of adaptation to extreme stress is universal; however, the actual manifestations may differ significantly cross-culturally and historically.

PART IV

Intervention

CHAPTER 10

PREVENTION

Until recently, the only form of mental health care available in the field of trauma was psychotherapy. When coping with extreme stress is considered as a normal process, however, there is a demand for lower level, yet methodically consistent methods of assistance. Not every victim needs specialized therapy, but most would benefit from some form of support and advice. Preventive intervention is the core concept here. By helping people in an early stage one can prevent serious disorders in a later period.

This chapter describes concepts which are helpful in designing preventive outreach programs for victims of traumatic stress. The first part summarizes the main elements of our theoretical approach to the process of coping. The second part analyzes the various aspect of psychological assistance following such events. We discuss psychological interventions and different components of intervention programs, and finally, we describe a specific program of psychological assistance that has been developed for employees of organizations who have fallen victim to acts of violence, such as bank robberies or hijackings. This includes the practical and organizational aspects of implementing preventive outreach programs in different settings.

10.1 Background

The study and the care of people after traumatic events, such as acts of violence, disaster and combat stress were for a very long time focused only on victims who suffered serious difficulties in coping with the event. The perspective for research and therapy was mainly psychiatric (e.g. Fromm-Reichmann, 1942). The effects of serious life events were considered as intense and prolonged for all people involved; psychiatrists were often astonished at the low rate of psychiatric casualties. This idea dominated the area of trauma research for a long time.

In the past two decades it has become clear that even under extreme stress the majority of people does not collapse or react in pathological ways. A central finding of studies on incidence and prevalence of posttraumatic stress disorders is that serious and prolonged disorders develop in a certain percentage of victims but certainly not in all. In fact, approximately 80% of all people who are confronted with traumatic experiences, work through them using their own resources and support from others. A minority of about 20 % struggles with stagnation in the process of coping (Card, 1987; Carey, 1977; Foeckler, Garrard, Williams, Thomas & Jones, 1978; Kilpatrick, Saunders, Veronen, Best & Von, 1987; Kleber & Brom, 1989).

This implies that most reactions to an extreme stressor are adaptive, normal and even necessary. People have to come to terms with the shocking event in one way or another.

This change leads to a new orientation in mental health care. Questions are now: Why do some people cope better than others? What are possible methods to prevent disorders? And who especially needs a preventive intervention? In this chapter we will try to answer these questions.

10.2 Summarizing theory on coping with extreme stress

A solid integration of theory is necessary as the basis for designing a preventive health care program. A general perspective on coping with drastic events has been developed, as we have shown in the chapters 8 and 9. This approach bridges the gap between psychodynamic ideas about the process of coping and the findings of the empirical research in the stress field.

The following elements are central to this theoretical approach concerning adaptation to trauma and are elaborated upon below:

1. The characteristic process of working through an experience of powerlessness and disruption.
2. The alternation between intrusion and denial.
3. The normal, functional character of the process.
4. The search for meaning as a motivating factor for the process.
5. The possibility of the occurrence of disorders.
6. The multiple determination of the content and intensity of coping.

1. *The process characteristic*

It is important to realize that the process of working through a serious experience extends over a certain period of time and that phases or elements may be distinguished within this process. The core of this process of coping consists of the handling of the experience of powerlessness, the disruption and the often very intense emotions which arise after the event. This process gradually has become well documented in scientific literature (chapters 4, 5 & 7).

2. *Intrusion and denial*

Two tendencies, in particular, are characteristic of the process of working through a serious event. *Denial* relates to an intra-psychic process, in which the implications of the event are denied in order to ward off the intense emotions. Examples of denial include: the avoidance of the place where the event took place, not wanting to talk or think about the event and emotional numbness. To outsiders it often looks like the person has recovered from the traumatic event.

Intrusion refers to the compulsive reexperiencing of feelings and ideas that directly or indirectly relate to the experience: nightmares about the event, startle reactions, preoccupation with the event, the wish to continually review the event and pangs of emotion.

These complementary tendencies of denial and intrusion alternate (Horowitz, 1976), although they can occur simultaneously on different levels. One moment the person trivializes the significance of the event to others, the next moment recollections flood the victim's mind, especially when finding himself in situations that resemble the original traumatic event.

3. *The process of coping is normal*

This element may seem trivial but it is implicitly denied by many authors. Even Horowitz uses the label 'stress response *syndrome*', implying illness. However, the alternation between intrusive reexperiencing and denial is useful. The individual experiences recollections in doses so that he or she is not overwhelmed by the intensity of emotions and other reactions evoked by the memories. The intrusive reexperiencing leads to a revision of the expectations and ideas of the individual.

In the literature on loss we find the idea of functionality of grief. Freud (1917) indicated that grief is not a pathological condition, but an adaptive process which does not require intervention. This is in contrast with some authors such as Engel (1961), who viewed grief as a disease. The individual breaks away from a loved person and this 'grief work' is meaningful. Averill (1979) and Bowlby (1980) even consider it a biologically functional process. Many characteristics of normal grief have a function, for example to evoke comforting behavior in others or to defend the individual against a too intense feeling of sorrow.

Grief is not only a normal and healthy reaction, but also a necessary one. People in one way or another have to come to terms with the death of an significant other. Avoiding grief leads to problems. 'To be able to mourn is to be able to change. To be unable to mourn, to deny changes carries great risk to the individual and to the (social) organization' (Pollock, 1977, cited by Volkan, 1979).

4. The search for meaning

Coping with traumatic life events can be considered an attempt to find a meaning in what has happened, as explained in chapter 8. This implies the search for an answer to the question 'why me?' One way to overcome the perceived chaos and to create order is to interpret the circumstances in such a way as to restore one's grip on the situation. An example of this is that people attribute an active role to themselves in circumstances in which they were actually completely powerless, like when sudden infant death occurred (De Frain & Ernst, 1978; Lowman, 1979). This is often referred to as cognitive or interpretive control (Thompson, 1981). The individual generates a sense of control and in this way diminishes the fear of a repeated occurrence.

People in general perceive the world to a certain extent as organized and comprehensible. One knows to some degree what to expect and for what reasons things happen. People live with internal certainties which are expressed in expectations, assumptions and ideas which regulate behavior and experience. An event like a serious traffic accident destroys these central intra-psychic structures (referred to as 'schemata'). An example of such a certainty is the idea, or rather the illusion of personal invulnerability (Wolfenstein, 1957). It is easy for people to participate in traffic because they assume, at whatever level of consciousness, that nothing will happen to them. Without such an illusion, many normal activities would evoke fear, hampering an individual's day to day life. After an accident there is suddenly a realistic sense of threat. One feels vulnerable. Could another accident happen any moment? An answer to the question 'why did this happen at this moment to me?' will help structure the individual's expectation of the future. The framework of expectations and assumptions has to be rebuilt in order for the person to avoid becoming tense and fearful again. This rebuilding is an essential element of the stress response.

5. *Disorders in coping with extreme stress*

Despite the fact that working through a traumatic experience is a normal process, people do suffer temporarily from symptoms. Although most people will struggle with some psychological and/or physical ailments they will handle these and recover without professional help.

Some victims, however, struggle with severe disorders as a consequence of the event they experienced. The stress responses in these people last much longer, get blocked or aggravated. The difference between normal and blocked processes of coping is not clear-cut. In general there is no difference in the nature of the reactions that occur, but rather in their intensity and frequency.

In the past years the term 'posttraumatic stress disorder' has become the label for these types of problems. As mentioned in chapter 2, posttraumatic stress disorders are not the only possible pathological response to traumatic events. Depressive reactions are also common. Dissociative reactions and brief psychotic episodes occur in some cases. The dynamics of coping are certainly discernible within these reactions. Surveying the studies which used representative samples of victims of traumatic events, we concluded that approximately 15-25% suffer from these disorders (Kleber & Brom, 1989). The criteria for the disorder, however, draw a somewhat arbitrary line between disturbed and normal processes.

6. *Multiple determination*

The intensity and content of the symptoms of coping, and in this sense also the occurrence of posttraumatic disorders, depends on several factors. This is the principle of *multiple determination of human behavior.* Different factors interplay to determine the social and psychological consequences of an event. This issue is discussed at length in chapter 9. The three general groups of determinants are:

1. The characteristics of the event and the context in which it took place (How real was the threat to one's life; was there material damage; when did it happen; how was the direct assistance; etc.).
2. The characteristics of the person and his personal history (Do experiences during youth predispose to certain ways of reacting to later events? Does one possess characteristics of personality such as rigidity and an external locus of control?).
3. The characteristics of the social situation. The extent to which people get support (in the form of encouragement, friendship and advise) from others in the nearest surroundings is especially important. Social support can moderate the impact of the event on the individual by influencing the appraisal of the event and by providing models of coping.

10.3 Psychological interventions

Our theoretical framework leads us to a new approach with regard to health care and assistance after serious life events (Brom & Kleber, 1989). The assumption of disorder loses its prominent place, and instead we emphasize that coping with extreme stress is generally a normal process, determined by various factors. Psychological interventions can be provided in different ways and at different times in the process.

We distinguish between three separate goals of assistance after serious life events:

1. stimulating a healthy process of coping;
2. early recognition of disorders;
3. psychotherapeutic treatment of posttraumatic stress disorders.

We will elaborate on the first two separately in this chapter. Different psychotherapeutic approaches are described in chapters 11, 12, and 13.

Stimulating a healthy process of coping

The methods of giving assistance directly after serious events corresponds in many respects to the practice of crisis intervention. A crisis is defined as a disturbed balance between a difficult and significant problem on the one hand and the available (internal and external) resources on the other

(Caplan, 1964). The experience of an overwhelming event is an outstanding example of a crisis. The impact of the event surpasses the capacity of the individual

Victims of events, such as acts of violence, are initially flooded by the impact of the event and the emotions which are elicited. The goal of crisis intervention is to return to a certain stability in the functioning of the individual. When people have sufficiently regained a feeling of control and tranquility, the efforts of crisis intervention mostly come to an end. However, from the perspective of coping as an alteration between denial and intrusion, the danger exists that assistance will be terminated prematurely in the case of a traumatic event. Immediately after the first phase of shock and bewilderment there is the phenomenon of 'outward adjustment' (chapter 5). The victim's denial of the emotional impact of his or her experience, gives the impression that he or she has completed coping and is doing fine. The fact that this happens in one of the first phases of coping indicates that the event still has to be worked through for the greater part. By breaking off the contact at this moment one misses the opportunity to follow the course of the process. Moreover, there is a risk that the support worker will reinforce the apparently adaptive behavior of the victim. This reduces the possibility for the prevention of serious disorders in the long run.

We have developed programs for stimulating a healthy process of coping with extreme stress, in projects for various categories of victims. One of these projects was a controlled outcome study of a special intervention program for victims of traffic accidents (Hofman, Kleber & Brom, 1990). Both from the findings of these research and intervention projects and from the results of other studies on counselling after serious life events (Raphael, 1977), we concluded that the following elements are important for the formulation of an adequate way of coping:

A. Practical help and information.
B. Support.
C. Reality testing.
D. Confrontation.
E. Several contacts with the support worker over a longer period of time.

A. Practical help and information

The importance of practical help is often underestimated. An example of this phenomenon was seen after the Xenia tornado in the U.S.A. in 1974 (Taylor,

1977). Experts and authorities expected that many survivors would need psychotherapeutic or other specialized help. The majority of the victims, however, initially struggled with practical problems: a diminished income, the lack of participation in the reconstruction, and difficulties with regular shopping. Immediately after the disaster there was a need for advice, referrals to institutions and other concrete support.

The same applies to other serious events. After acts of violence and traffic accidents, the support worker should give practical information on the procedures which follow. In the case of specific problems referrals to specialized institutions should be made. Medical questions often arise and require elucidation and referral. It is important that mental health professionals are well aware of the possibilities to obtain different forms of practical help.

General information about the process of coping and accompanying symptoms can also be very reassuring to victims, who are mostly unaware of the consequences they may experience. The support worker should explain that such an event is followed by a process of coping and that this process is accompanied by various psychological and physical reactions. It is of utmost importance that the information is given in the right way (e.g. in close connection with the experiences of the victim) and at the right moments (e.g. not during the first hours after an assault when, in general, victims are not open to new information). The information enables the victim to put his own reactions into the framework of normal coping. In this way secondary symptoms, such as worries over physical or psychological symptoms, which occur frequently, can be prevented. Finally, it is important to inform the victim about the help he can expect and the way in which it will be offered to him.

B. Support

Rendering support is by far the most important task in the assistance after serious events. By this we mean the creation of a safe and quiet environment, which enables people to realize that the event really is over. This condition is necessary in order to allow the coping process to begin. An accepting attitude and a genuine interest in the fate of the victim are indispensable to the support worker.

Support, however, is a hazy expression (chapter 9) and it is necessary to define it with a description of the specific activities of the support worker. In the first phase of contact, the support worker tries to bring structure to the experiences of the victim. In the individual interview setting, this is done by first *listening closely* to the victim and then *labelling* the emotion which is expressed by the victim. The worker directs the victim only in a limited way to explore his or her feelings further. In general, any intense discussion of emotions in the first phase after a severe event increases the confusion and accentuates the crisis.

The help of support groups is important, but can, of course, not replace the support of people's own social network. With this in mind, the program tries to stimulate victims to mobilize their personal network and to draw attention to his or her situation and feelings. But listening to traumatic experiences is difficult and frightening. The lack of attention given to victims by the people in their surroundings is understandable. People need a certain amount of self-protection and can hardly bring themselves to genuinely listen to the terrible stories the victims tell us. This makes it even more crucial to train victim support workers to be able to endure this task of lingering on the details of these experiences.

As time passes by, the victim will need this structured form of support less and less. The worker will then adopt a more firm and confronting attitude. Reality testing and confrontation are important aims ot this stage of the intervention, although an atmosphere of understanding and trust remains the most important basis of the relationship.

C. *Reality testing*

As mentioned before, coping with a traumatic event can be considered as a process of forgetting and retrieving; this is for the most part automatic and unconscious. Many victims feel that they have lost or forgotten part of what they went through, and this is one of the reasons they want to go over it again and again, trying to reconstruct the event. Often they realize after some time that the event was slightly different than their first impression of it. These are important and healing processes.

A second form of reality testing is necessary because of the uncertainty about the symptoms which people experience. The difficulty in concentration and the forgetfulness which frequently arise and can last for a long period are particularly frightening for many people. Discussing these and other reactions as a normal part of the experience generally changes the appraisal of the symptoms. They become less alien and frightening.

D. Confrontation with the experience

Denial and intrusive reexperiencing are two crucial poles of the process of coping (chapter 2 and 8). Denial in itself is a functional mechanism, preventing people from being overwhelmed by their emotions. If, however, there is no alternation between denial and intrusion and the victim fully suppresses the unpleasant feelings, the chance increases that coping will not be successful and disorders will appear in the long run.

In the later stages of the contact the support worker takes the victim back to his recollections of the event and discusses the course of it in detail. The goal of this is to counteract the avoidance tendencies. This also diminishes the chance that the reaction of the victim would become interwoven with longer existing patterns of behavior. An active attitude of the support worker is required to bring about this confrontation.

E. Several contacts over a longer period

Intervention programs should include several interviews the last of which is at least two to three months after the event. As mentioned before, the approach of the interviewer changes over the course of time. In this way he or she is able to follow the course of coping sufficiently and to recognize disorders in an early phase.

In the first contact with the victim, the support worker's focus is mainly on getting acquainted, on supportive behavior and on giving information. The support worker's attitude is understanding, empathic and structuring. The goal is to give the victim back as much as possible his or her sense of control. Later on, the confrontation with the event becomes more important than the support and information giving.

Early recognition of disorders

The second aim of psychological assistance after serious life events is early recognition of disorder in the process of coping. This presupposes knowledge about the relation between signs in an early phase and the occurrence of disorders. This knowledge, however, is limited at present. For a better understanding of the factors that influence the occurrence of disorders we refer to the principle of multiple determination mentioned above. The characteristics of the event, those of the victim and those of the social situation interact in their influence on the coping process.

It is important to detect risk factors in the individual victims in order to create a global expectation of the way people will handle the situation. From research we know that the more serious the event, the greater the risk of disorder. But also, the stronger the sense of personal threat, the greater the risk. In this way we can indicate a number of risk factors. We must realize, however, that these factors are the result of research on groups and can never give precise prediction about individuals.

An important issue for the practice of victim support is the relation between severe emotional reactions in the early phases of coping and the severity of later reactions. If a victim comes to us fully upset, is this a good sign because he or she is expressing his or her feelings? Or is it a sign that his or her capacity is overtaxed and that therefore we should count on lasting problems. Few systematic evaluations on this subject are available. Our impression, however, is that strong emotional reactions in an early phase are positively correlated with the severity of later reactions. This runs counter to some general expectations, but seems most in agreement with the currently available research data (Van der Ploeg & Kleijn, 1989; Weisaeth, 1989b).

The identification of risk factors can be made productive in two ways; on the one hand to determine target groups who need special attention and on the other hand to give a basis to the policy for individual assistance. This will be a very important topic for future research.

Group versus individual assistance

There is a discussion among clinicians about the relative efficacy of group versus individual assistance after traumatic events (Van der Kolk, 1987). This discussion parallels a larger one about the comparison between group and individual psychotherapy. Group intervention has the advantage of the authority of the group, which is often much stronger than that of an individual counselor. Individual counseling has the advantage of a custom-tailored intervention, which allows more adjustment to individual needs.

The same applies for the reentry into society of victims. In chapter 12 this aspect of trauma will be outlined in connection to the concepts of self-psychology. On the one hand a group can be an important practice ground for victims for communicating with others about their experiences. On the other hand there is the danger that a group of victims develops a pseudo-identity, based upon being victims. One advantage of individual counselling is that there is plenty possibility to test the individual functioning and observe early signs of disorders. This might be more difficult in a group setting.

In the end, the decision about whether to offer group or individual counseling will probably be directed by practical circumstances. When larger numbers of people are involved in disasters or other traumatic events, practical considerations may lead to the decision to employ group counselling. When fewer people are involved and the manpower is available, many people will prefer individual counselling. In later stages of the coping process, when a more probing confrontation with the traumatic event is needed for a preventative intervention, individual counselling will most surely fit the task.

Other models of intervention

In the past years several other forms of intervention after severely distressing events have appeared. Some of them became popular, despite a weak theoretical basis. We will discuss two approaches, that are different from our model, as well as the role of self-help groups.

In the early eighties Mitchell (1986) developed a program under the name of *'Critical Incident Stress Debriefing'*. Without supplying any theoretical background, the program gives clear guidelines for setting up group sessions after traumatic events. The program is mostly used in police and other emergency services, where groups of people go through severely distressing experiences. Debriefing is described as a psychological and educational group process. The main element of the program is a debriefing session which is held 24 to 72 hours after the event. In this session practical information about what happened and about expected symptoms is provided and people are given the opportunity to express feelings and ask questions. This, in possible combination with a so-called 'defusion' session immediately after the event, is intended to prevent posttraumatic stress disorders.

Critical incident stress debriefing contributed to the care of emergency service workers, especially because of its appeal and its easy applicability. It must be clear, however, that the debriefing model has some serious shortcomings. The program is solely aimed at reducing stress reactions. The total disregard of aspects of coping, heightens the risk of attaining a quick return to normal functioning, without appropriate working through of the experiences. Furthermore, the question arises: what is the effectiveness of group intervention in early stages of the coping process? How do we prevent individuals from slipping through our fingers when we want to assess early signs of disorders? This is especially relevant in the light of the knowledge (Weisaeth, 1989c) that people, who are reluctant to participate in intervention programs, belong to a high risk group. In our opinion, any intervention model should deal with these issues on the basis of a consistent theoretical framework, which is absent in the work of Mitchell.

Another appealing idea, which often comes up when dealing with organizations that are prone to suffer from extreme stress events, is the possibility to prepare people for traumatic events even before their occurrence. Weisaeth (1989a) found in his study of industrial employees that adaptive behavior in a disaster setting was predicted by the general level of disaster training and experience.

Stress inoculation training was developed in the seventies by Meichenbaum (1977). Its aim is to prepare people for the stresses of life, decreasing feelings of helplessness and promoting active ways of coping with stress. Stress inoculation consists of a combination of recommendations, warnings and training, which 'inoculate' against the emotional and material implications of the expected event (Novaco, Cook & Sarason, 1983). Such an intervention model was used by Veronen and Kilpatrick (1983) to deal with the effects of rape. The question is whether this model is applicable *before* distressing events, in order to improve the ability of people to cope. In this respect we could speak about 'trauma inoculation'.

Without going into this question deeply, we will indicate some possibilities and limitations. Trauma inoculation, according to the principles of Meichenbaum, but specifically attuned to trauma, can facilitate coping after events with a limited impact. Training bank employees to deal with robberies, so they know what to expect and what to do both during and after the event, will help in the majority of robberies, because they do not usually end in bloodshed. However, when there is a personal confrontation with death and violence, the usefulness of prior training will be minimal. There is no remedy for fear of being killed, for the grim confrontation with death or for the shock of the loss of a close person.

Finally, we should draw attention to *self-help groups*. There are self-help groups in many domains of human functioning. It seems a very natural thing for people who went through similar experiences to support one another. Self-help groups provide more formal and specific communication than other interactions, for example, with relatives. Again, this kind of help can be very powerful. Group interaction will both elicit strong responses and give feedback to the individual, which moderates emotional responses.

Self-help groups in general only come into being when there is some societal acceptance of the suffering. This is apparent when we look at two examples: for parents who have lost a child, there are many support groups all over the world, while victims of traffic accidents have hardly organized themselves. What seems to count here is not the frequency of the occurrence of the traumatic events, but the acceptance in the environment that this event needs processing.

Self-help groups are a very valuable tool for communicating about painful experiences. Undoubtedly they should be promoted and supported. Not only can they have a positive influence on individual adjustment, but they also stimulate social action in the specific field of victimization. These activities of self-help groups have a very broad influence and encourage victims to share their experiences with others, even if not in the framework of the group.

10.4 Care of victims of violence in organizations

In several professions there is a risk to become the victim of violence. Banks have always been the target of robberies, hold-ups, kidnappings, and blackmail. More and more organizations, however, are been being confronted with work related violence, partly because of the increase in crime. Examples are: employees of department stores and supermarkets confronted with hold-ups and hijackings, gas station personnel, police officers involved in shooting incidents and accidents (Gersons, 1989), personnel of security companies, prison guards and penitentiary officers. These employees have to cope with the aftermath of life-threatening events in their jobs, yet these events affect not only the individual, but also the organization, resulting in job tensions, absenteeism and low morale. Therefore, employees, personnel workers, and managers are realizing that assistance is indispensable.

During the eighties we developed an intervention model in The Netherlands to assist victims of work related violence. The model has been specified and applied in standardized programs for different organizations whose employees have been robbed or kidnapped. In this section we describe general features of the model.

Principles of implementation of intervention in organizations

Mental health care of victims within an organization can only successfully implemented when some guiding principles are followed. The authors

(Kleber & Brom, 1986) formulated these guidelines during the development of intervention programs for several large banking organizations. The principles are summarized as follows:

a. All personnel involved participates in the intervention.
b. Assistance must be provided directly after a serious event.
c. There should be several contacts over a longer period of time (preferably a couple of months) between the victim(s) and the support worker.
d. The management assigns a staff member who is in charge of the assistance of the victims.
e. All procedures are set down in an explicit scenario.
f. Certain changes in job conditions should be permitted, such as special procedures in case a person needs a leave of absence.
g. Referral to psychotherapists in case of serious disorders.
h. Continuous evaluation of the intervention program is necessary, especially in the first period after implementation, in order to observe the organizational and the professional adaptations which have to be made to fit the specific needs of the organization.

Let us go into some of the above guidelines. The best way to implement assistance of employees who have been robbed or kidnapped is standardization of intervention. This implies that all personnel involved participates, unless they actively object. Standard procedures have the advantage that stigmatization is avoided. Moreover, high-risk groups of victims are more easily reached, because everybody takes part in them. One of the risk factors is the strong avoidance of the memories which leads some victims to turn down a noncommittal offer of assistance.

This kind of preventive mental health care has an active or 'outreaching' character. An approach of 'immediacy, proximity and expectancy' (chapter 3) has found to be more adequate in victim support programs than a passive attitude. Of course, such an active approach is only possible from a perspective that emphasizes normal coping and not disturbances. The responsibility and independence of the person should be emphasized.

Assistance should be formulated as an official organization policy. Immediately after serious events both management and police have priorities that are not always in line with the interest of the victimized employees. The reactions of the victims are often overlooked. Explicit procedures and clear responsibilities for the care of the employees will prevent victims from being 'forgotten'.

It is recommended that specific staff members will be appointed to be in charge of the assistance of the victims. Preferably they should be skilled in counselling and supporting crime victims. These persons must be available to assist the employees immediately after the robbery or kidnapping and must also conduct the follow-up interviews. It is suggested that the personnel counsellors do not have a direct association with the career planning of the victimized employees; the victims should feel free to express their feelings and thoughts to the staff members (social worker, industrial psychologist or specialized personnel worker). The counsellor should not be forced to report on the victims. His or her function is to support the victims, even if sometimes this is not in line with the interests of the organization. Training of organizational victim assistance workers is an essential part of the intervention model, although practical experience has shown that specific skills in dealing with victims of violence are not easily acquired by the trainees. Supervision of counselors and maintenance of intervention skills, therefore, are essential.

An advantage of this kind of intervention is that assistance is being delivered by people within the company. This health care is therefore on the crossroad between professional help and layman help. If everything is arranged properly, outside professional psychologists are only involved in cases of victims who need more specialized help.

Finally, organizations whose personnel could be victimized by job-related acts of violence must develop procedures for consequences such as the temporary absence of an employee because of the traumatic event. The label 'sick leave' creates stigmatization and could interfere with the normal process of coping, in particular when this absence occurs in the first weeks after the incident. Another work related issue is the transfer of the employee to another position in the organization. There is always a chance that a victimized employee cannot cope with traumatic experience and therefore cannot continue with his or her work. Job transfer or job rotation can be adequate solutions.

Specific procedures are also necessary for employees who have experienced extreme events repetitively within a short period of time: for example, a bank cashier left her job after 13 years, after she was confronted with the fifth hold-up, the third in the last year. Some institutions, such as those in high crime areas, are the target of far more violence than others. Being repeatedly victimized is an important risk factor for developing coping disorders. The personal sense of security and invulnerability is shattered by the repetition of events. It is often impossible for these individuals to return to their job situations. Job transfer may be required for the employee and/or a referral to a therapist specialized in traumatic stress.

In repeated experiences of similar traumatic events the impact of psychological help is clearly limited. There is a limit to what people can cope with. Working with a high risk population for retraumatization one should be aware of these limitations and not overestimate the possibilities of counselling. Sometimes the advice to leave a dangerous environment is the only justified advice from a psychological perspective.

Elements of organizational climate

If everything is implemented properly, organizations could run their own intervention programs. Assistance by outside professionals would only be necessary in the case of coping disorders, special problems, and referrals to specialists. This, of course, is not the situation. Intervention programs are fairly new, and are still in the stage of introduction. In the course of development and implementation some shortcomings will be discovered in the programs, so attention must be paid to new issues, as they arise.

The organizational context is extremely important. The impact of organizational setting is often underestimated. A concrete intervention program is never an easy derivative of a theoretical model. Many modifications are required in order to adjust to the needs of a given organization.

Organizations have their own atmosphere, often characterized by no-nonsense thinking and a reluctance to spend money on psychological issues. 'As long as one does not talk about stress reactions, they will not experience them' is often the prevailing attitude in organizations. Another conspicuous aspect of the organizational culture is an emphasis on macho behavior, sometimes labelled as the 'John Wayne' culture. The attitude of 'stiff upper lip' or 'being a tough guy' pervades many organizations in which workers are confronted with violence or accidents. In these organizations notions such as 'violence goes with the job' or 'if you can't stand the heat, get out of the kitchen' are often expressed. These attitudes are common in police departments, emergency personnel, fire fighters and public transport personnel. Colleagues expressing emotional distress after acts of violence are sometimes publicly denounced as 'sissy' or 'softie'.

Individuals in these kinds of organizations frequently keep their contacts with support workers a secret, often with good reasons. The following is an example of this culture: a penitentiary officer with 14 years of service developed posttraumatic stress disorder after being taken hostage and injured in an escape attempt. Considerable attention had to be devoted to feelings of guilt and shame in treatment. During the years this officer had often derided colleagues who expressed emotional distress after violent events. After being victimized himself he realized how much suffering he must have caused with his remarks.

Despite the avoidance of stigmatization by using a standard approach of health care and despite an emphasis on coping with extreme stress as a normal process, there is still a strong resistance against psychological help. A macho culture is still prevalent in many organizations and as a result victims are often neglected in a harmful way.

10.5 Conclusions

Since the eighties the interest in supporting people after traumatic life events has increased. More and more psychological assistance is developing in the direction of preventive counselling, and more and more the accent will be on early recognition of disorders. The determination of risk factors and their relative influence will be useful tools for planning and implementing mental health care.

In particular, interventions are prompted by a growing attention to work related traumas. Employees of organizations are the target of robberies and other types of crime. Although controlled outcome studies have not been conducted, the programs seem to work well and are being appreciated by the victims.

There is a pressing need for outcome studies of the different intervention models that are available. By looking at the research on psychotherapy outcome (chapter 14) we may suspect that there are no differences between the models, as far as outcome is concerned. Although the investment in counselling programs has grown tremendously, there has not been a demand for research to show its efficiency. In this field people seem to rely on the positive feedback from the victims. In the future we would strongly suggest research that would further specify the intervention models and would test their applicability and effects.

CHAPTER 11

BEHAVIOR THERAPY

Behavior therapy has been late in recognizing and dealing with the impact of severely distressing events. In the seventies, some case reports on behavioral treatment after trauma were published (Blanchard & Abel, 1976; Kipper, 1977), but these referred mainly to the treatment of isolated complaints. Later, in the eighties, articles appeared that dealt with the process of coping with traumatic stress as the central issue. Theoretical views on posttraumatic disorders emerged and in line with these several treatment modalities were developed. Since then the field has grown quickly and procedures such as flooding, desensitization and cognitive restructuring are being used in the treatment of posttraumatic stress disorders (Peterson et al., 1991).

The development of behavioral theory on coping with trauma appears to be in accordance with the general state of the art in the field of behavior therapy, i.e. two lines of thought prevail. On the one hand a conditioning model is proposed, accentuating an interaction between essentially simple learning procedures. On the other hand, cognitive learning theory is being used to explain the phenomena after traumatic events.

In this chapter we will first present current behavioral theory on the effects of traumatic events. Second, we will outline the technique of trauma desensitization, a behavioral therapeutic method for excessive reactions to traumatic events.

11.1 Learning theory and coping with traumatic experiences

In chapters 3 to 7 we described the consequences of traumatic experiences. The core of posttraumatic reactions is the development of a specific conglomerate of symptoms after an experience of disruption and helplessness. Can the development of these symptoms be explained by means of learning theory? Which learning processes are at the basis of behavior after traumatic experiences and which processes can be held responsible for the decrease, continuation or even increase of these reactions?

In this section we focus on two models that address the questions posed above. This is, of course, a limited choice of the models that have been developed in behavior therapy. It is our contention that these two models, which are widely cited in the recent literature on traumatic experiences (Keane, Fairbank, Caddell et al., 1985; McCann & Pearlman, 1990), are the most insightful and therefore the most useful for clinical practice.

The two factor model

The two factor model (Mowrer, 1960) assumes two forms of conditioning, classical and operant, at the basis of the phenomena after traumatic experiences. The emotional reaction to a traumatic event can be understood through classical conditioning (Kolb & Mutalipassi, 1982). There is an unconditioned stimulus that triggers an unconditioned response. In other words, a traumatic experience results in an instinctive startle response. According to the principles of classical conditioning, all kinds of stimuli, which were present in the original experience, may become associated with the startle response. These stimuli will then be able to elicit responses that are similar to the initial response. However, if these conditioned stimuli occur without the unconditioned stimulus, extinction will take place.

In order to illustrate this process of conditioning and the subsequent extinction, we will present the story of V, who became the victim of a hold-up.

> *Twenty-one-year-old V has been employed as an administrator at a bank for the past few months. Four months before he contacted us he had been asked to substitute as co-driver in an armored car. The car was to drive from a bank to the office where V worked. When the vehicle attempted to drive into the office parking lot, two masked motorcyclists began to shoot at the chauffeur. V was looking the motorcyclists directly in the eyes when he felt he had been hit in the arm. In a total panic, but alert enough to avoid*

any further shots, V fled inside the bank, where he was received by a colleague. It was at that moment, according to V, that he first felt pain in his arm. When the chauffeur ran into the building moments later, V flew into a panic again. He thought he was looking into the eyes of the robbers again.

V's life has changed since this experience, partly due to the partial paralysis he suffered as a result of the shot in his arm. During the first few weeks he was easily startled. Minor occurrences, such as accidentally dropping a hammer, were capable of frightening him and causing him to burst out crying. During this period of time he tried to avoid thinking about the experience. When he did remember it, he quickly became upset. After some four weeks, the startle reactions did no longer occur after minor stimuli, but were triggered by passing motorcyclists and when someone to his left suddenly made a movement. V began to dream during this period of time and would wake up in a panic without remembering the content of the dreams. Some weeks later, his dreams became clearer and he pictured his assailants. V felt that the images became sharper and more complete in time, whereas the feelings of panic decreased in intensity. After 3.5 months he could picture the entire event as in a dream; he felt anxiety but not panic.

During the first two months V talked a great deal about the event. He felt compelled to tell the story over and over again. Fortunately, the people around him lovingly gave him the opportunity to do so. After 3.5 months, this need, as well as his tendency to avoid the scene of the robbery, were gone, although the sight of the armored car still triggered emotions.

With the help of this particular example we are able to clarify the conditioning process of an anxiety response. The enormous fright evoked by the potentially fatal situation appears to attach itself to a number of characteristics of the original event. Accordingly, V is initially frightened by almost every change which occurs in his environment. In particular, sudden loud noises make his fear emerge. In terms of conditioning, many events, especially unexpected ones, function as conditioned stimuli.

Before long, the reactions diminish, and the similarity between a stimulus and the original event will need to be greater in order for an anxiety reaction to take place. The link between conditioned stimulus and response is broken. According to classical conditioning theory, this extinction occurs during confrontation with conditioned stimuli in the absence of the unconditioned stimulus.

In a limited number of people, however, there is a conditioning of such reactions without extinction. The intense anxiety reactions and other emotional consequences persist or become triggered by even more situations than was the case in the beginning. Referring back to Mr. V, his sensitivity to sudden noises may have increased and his tendency to avoid thoughts and places that remind him of the attack may have become stronger. The latter process can be explained by so called 'higher order conditioning', in which stimuli, present preceding or during the traumatic event, acquire the same anxiety-eliciting characteristics. Mr. V might have been in a drugstore when feeling terrified by the sound of a passing motorcycle, which might have led him to avoid going to this drugstore again. Another phenomenon is 'stimulus generalization', through which one stimulus that resembles another stimulus can evoke the same response.

The persistence of certain reactions may be explained in terms of avoidance behavior, which in itself becomes a part of a conditioning process, in this case operant conditioning. Avoiding the location of a traumatic event, for example, is reinforced because the negative emotional experience elicited by this place does not occur. When avoidance behavior is rewarded by the environment and when unpleasant events do not occur as a result, the chances are that this behavior will persist. With such continuous avoidance behavior, the extinction of emotional reactions is prevented.

In summary, two factor learning theory maintains that there are two active ingredients in trauma conditioning. The first is the emotional part, where extinction has not taken place, and the second is the factor which inhibited the extinction by preventing the confrontation with the most emotionally charged stimuli.

The two factor model offers two approaches in regard to the treatment of people who are not able to cope with traumatic experiences. In the direct approach, the goal is the advancement of the process of extinction that has not yet occurred. The most important is the gradual, or even forced, confrontation of the person with the conditioned stimuli. When this happens to a sufficient degree, extinction will occur and the feelings of anxiety as well as the avoidance behavior linked to it will disappear. The direct approach aiming at extinction is the basic principle of both the flooding technique and desensitization.

A second possibility for treatment focuses not so much on the 'classically conditioned' responses, but on the behavior that obstructs their extinction. This method does not aim at achieving extinction through direct confrontation, but rather at creating a situation in which the individual is able to resume the 'normal' course of a coping process.

The factors responsible for the failure of extinction are very diverse.

Personality traits, such as a rigid need to control one's emotional reactions, as well as responses from the environment, which often supports a non-expressive attitude, are important in coping. In addition, certain reactions can be 'over-conditioned', as is the case when people have gone through several subsequent traumatic experiences. In such cases the emotional reactions often are so strong that the confrontation with one important stimulus means the avoidance of another. The focus of the intervention will obviously have to be chosen according to the nature of the problems.

A related conception of posttraumatic stress disorder within learning theory is the formulation of Pitman (1988). Posttraumatic symptomatology is put into the framework of network theory, which contends that emotional responses are part of an information structure in memory. Network theory may more easily account for the occurrence of responses, such as flashbacks, to stimuli, which have no obvious connection with the response. Pitman also arrives at the conclusion that exposure is the treatment technique of choice. His formulation stresses that the therapist should be aware of a broad range of stimuli that play a role in the conditioning process after a traumatic event.

The model of learned helplessness

A second relevant model, that emerged from learning theory is that of 'learned helplessness' (Seligman, 1975). In the sixties, Seligman, an experimental and clinical psychologist, studied the acquisition of avoidance behavior. He administered a number of electrical shocks to two groups of dogs in an experimental setup in which they had the possibility to avoid the shocks (see chapter 8). One group had prior experience with an experimental situation in which they could not avoid electrical shocks, the second did not have prior experience of this kind. A dog that is not given shocks beforehand generally adopts effective avoidance behavior rather quickly. Much to Seligman's amazement, however, dogs that were exposed first to unavoidable shocks, afterwards did not learn efficient avoidance behavior as easily. It appeared the dogs had learned that the shocks were unavoidable, and had resigned themselves to that fact.

Seligman concluded that learning does not involve the establishment of simple stimulus-response connections. He proposed a model of cognitive learning, i.e. the acquisition of knowledge or expectations of the effectiveness of one's own behavior. The dogs thus became helpless because they had learned that shocks occurred independently of their own behavior. Seligman referred to this phenomenon as 'learned helplessness'. He argued that it was not only present in animals, but that it was a central element in the development of depression in human beings.

Since Seligman's experiments, much research has been done into the behavior of animals and humans in situations that evoke helplessness. A revision of the model was proposed by Abramson, Seligman and Teasdale (1978) in order to solve inconsistencies. In the original model, for example, it was argued that learned helplessness develops after events that occur independently of the person's behavior. No distinction was made between the consequences of an event that might have been influenced by someone, and the consequences of an event that could not be influenced by anyone. When we consider car accidents, for example, there is a large difference between coping with an accident, that could have been avoided through adequate behavior, and an accident that clearly was impossible to prevent.

The revised model deals with this and other shortcomings by using attribution theory, which presupposes a relationship between the source to which events are attributed and the emotional situation in which one finds oneself. For example, a person's self-image in the aftermath of a traffic accident is determined, for a large part, by the degree of helplessness he experiences. Those who feel they could have avoided the accident attribute it to themselves (internal attribution) and will have low self-esteem. Their condition is one of 'personal helplessness'. Those involved in an accident in which they perceive that neither they themselves nor anyone else in the same position could have changed the situation, will attribute it to external causes (external attribution). Their self-esteem is not changed by this event, and this condition is described as 'universal helplessness'.

In the revised model Seligman and colleagues used attribution theory to specify the nature of cognitive attributions. An internal, stable, general attribution, in their terminology, is most often associated with depression. An example of such an attribution after an act of violence is:

'I was mugged because I have always been a naive person'.

In contrast, an external, instable, specific attribution is generally connected with better ability to cope. An example of this:

'This mugging hit me for no reason, i.e. the mugger happened to be where I was at that time and he was looking for a passerby'.

Traumatic experiences are pre-eminently uncontrollable. No defense is possible, and the person is flooded with very unpleasant experiences. According to the model, this event will cause helplessness. The person experiences a lack of control over the events in his life. This is different from the feeling of having caused a event through one's own behavior. The element of uncontrollability is then joined by fear for one's own actions. The person blames himself for the negative event (internal attribution).

In general, little is known about the circumstances in which people blame global or specific factors for a traumatic event, i.e. whether they attribute the event to a global factor such as fate or personality or to a specific factor such as the unique characteristics of the event. Similarly, little can be said about attribution to either stable or unstable factors. The conditions under which such attributions are made need further specification.

Learned helplessness and its related phenomena can no longer be explained by the situation itself. Factors such as the individual style of attribution, the way in which one has learned to evaluate one's own part in the event, now plays a role. For example, what conclusions do we draw from the fact that we become a victim through a violent crime? Everyone is obviously afraid that such an incident might occur again. But for someone who has learned that 'one does not become a victim of such a crime for no reason', it is probably far more difficult to attribute factors in a way that enables him to live on without fear. In clinical cases we usually see general and stable attributions. Clients feel overwhelmed and feel they are not strong enough to deal with certain aspects of life. The individual learning history of the client, i.e. how he has learned to see himself and how he has learned to draw conclusions based on events in his life, plays a major role.

Based on the perspective of learned helplessness, we distinguish a number of goals for the treatment of disorders in the process of coping with traumatic experiences. The most important one is restoring the client's feeling of control. The victim has lost the feeling that he can influence the course of events. Regaining this sense of control, either by cognitive or by behavioral means, is a main task for the therapy.

The appraisal of the traumatic experience has taken a central place in the world of the patient. Diminishing its influence within the perception of the patient is another major task. He should be made to realize that the occurrence of a traumatic experience does not imply that it will happen again (converting stable attribution into unstable attribution) and similarly that no other or additional misery will follow (converting global attribution into specific attribution). These therapeutic tasks are essentially implicit elements of most behavioral treatment modalities for disorders in coping with extreme stress.

11.2 Trauma desensitization

In the remainder of this chapter we will elaborate on theory and application of trauma desensitization. The first ideas on desensitization can be found in the third decade of this century in the well-known experiment with 'little Albert', an 11-month-old child. Although this research is objectionable on ethical grounds, it is seen as a major contribution to the development of ideas on the learning of fear. Researchers Watson and Rayner (1920) were interested in learning processes. During the experimental phase of the research project, little Albert played with a tame white rat while the researchers hit an iron bar with a hammer. After Albert was frightened seven times by the noise, it became evident he had become afraid of the white rat. In addition, this fear grew to include other things that resembled the white rat. A fur coat, woolly fabrics and rabbits could now also evoke an anxiety reaction in the child, even though they were not included in the experiment.

This study and the development of ideas on learning anxiety reactions were continued in the work of M.C. Jones (1924). Jones worked with three-year-old Peter, a boy with a phobic fear of white rats, rabbits, fur coats, etc. She treated Peter by bringing a rabbit increasingly closer to him during each session. Over the course of 80 sessions, Peter learned to play with the rabbit without fear. His fear for similar objects disappeared. The principles involved in learning or overcoming fear, which were developed in the work of Watson and Jones, were not elaborated upon until much later, when they served as the basis for the development of one of the most important methods of behavioral therapy, that of systematic desensitization.

In 1958, Wolpe was the first to formulate the principles of systematic desensitization on paper. The main elements of this method of treatment are:

- the formulation of a hierarchy of fear-evoking stimuli;
- the learning of relaxation techniques;
- the confrontation of the patient with anxiety-evoking stimuli while in a relaxed state.

The stimuli consist not only of real objects, but also of ideas and notions the patient has about certain objects or situations. In the former case we use the term desensitization in vivo, in the latter we speak of desensitization in vitro. Desensitization is generally applied as a technique for reducing anxiety. The method is also used, however, in dealing with problems, such as speech disorders, insomnia, nightmares, sexual impotence, chronic alcoholism, and various other complaints.

There has been some controversy between proponents of desensitization and proponents of exposure techniques without relaxation. Desensitization was seen at one end of a continuum and flooding, i.e. the continuous confrontation of the client with anxiety eliciting stimuli, at the other end. Flooding seems to be favored by many authors in the field of trauma because of the vehement responses which are associated with traumatic experiences. In the past years there seems to be a blurring of the controversy and the flooding method of Keane et al. (1985) includes relaxation and resembles in many ways the method we will describe below.

In recent years, there has been increasing interest within the circles of behavioral therapy in the process of coping with traumatic experiences and related problems (Fairbank & Brown, 1987). Research and theory formation, for instance, developed in the area of coping with rape (Holmes & Lawrence, 1983; Kilpatrick, Resick & Veronen, 1981). The main area of application of trauma-related methods derived from behavioral therapy, however, is war stress. Flooding, in particular, is being applied widely in recent years to war veterans (Boudewyns, Hyer, Woods et al., 1990; Fairbank & Keane, 1982; Fairbank, Keane & Malloy, 1983; Keane & Kaloupek, 1982; McCaffrey & Fairbank, 1985), although systematic desensitization is also presented as a possibility (Bowen & Lambert, 1986).

11.3 Explanations of the effect of desensitization

Wolpe (1958) designed the method of systematic desensitization as a 'counter-conditioning' technique. He presupposed a process in which a certain reaction (anxiety) is made impossible by the occurrence of another reaction (relaxation). Thus desensitization was designed to break classically conditioned stimulus response chains.

Since Wolpe's assertions much research has been done on the effects and the theoretical background of systematic desensitization. Little doubt exists concerning the method's effectiveness. Time and again it was shown that positive results could be achieved, even in comparison with control groups (Gottman & Markman, 1978; Murray & Jacobson, 1978)

The matter becomes much more problematic, however, when we consider the question of which ingredients of the desensitization procedure are instrumental in achieving the effects. Closely related to this is the question of theoretical principles on which the procedure is based. Many researchers have attempted to solve this problem by leaving out one of the components of the procedure, for example, the relaxation or the hierarchy of anxiety-evoking stimuli, and subsequently reassessing the effects. The only conclusion from all this research is that no single component can be held responsible for the effects found. On the basis of the findings, which indicate the effectiveness of desensitization, but without pinpointing any specific factor, that can be held responsible for the effect, Yates (1975) concludes that 'systematic desensitization is left with only its smile'.

Reviewing the literature dealing with the explanation of desensitization (e.g. Van Egeren, 1971), it becomes clear that notions on the effect of therapy were not just highly mechanistic but also simplistic. It was as if the complex interaction between therapist and client that occurs during the process of desensitization should have been explainable in terms of one principle.

In view of Yates' disappointed conclusion, it would be more productive to take into consideration the complexity of the interaction between therapist and client. Lick and Bootzin (1975) concluded that the expectations of the client are an important variable in the effect of the treatment. This has led people to depreciate the actual procedures of desensitization. Their conclusions, however, could also lead us to look for additional procedures to boost the expectations of the client. Similarly, the perspective of Wilkins (1971) may be used to clarify and emphasize practical aspects of desensitization. Wilkins quotes research that indicates that providing the client with information about his behavior and other reactions is therapeutically beneficial. The combination of this view and the indications given by Averill (1979b) about the stress-reducing effect of information about a stimulus or an expected reaction to it, should be used for the further refining of the procedure. This may lead to suggestions as to how the therapist can increase his effectiveness through giving information and teaching the client to distill information from his own reactions.

The plenitude of research on desensitization (Kazdin & Wilcoxon, 1976) has clearly shown that the question of finding the sole explanatory principle is not a useful one. Despite the lack of clarity that exists about the theoretical background of desensitization (Evans, 1973), we must ask ourselves to what extent desensitization as a method of treatment relates to the process of coping with traumatic experiences. In order to answer this question, we consider two theories used to explain the effect of desensitization.

Theory serves two purposes in the realm of psychotherapy. On the one hand it is supposed to give a scientific backing to complex human interaction. But secondly, and for clinical practice this may be much more important, theory should guide the therapist in deciding about aims and means in his daily work. The scientific basis of the theories we will discuss, cannot be considered to be completely adequate. They do, however, provide a stable framework for systematic clinical work.

Counter-conditioning

As we have described above, desensitization was viewed by Wolpe as a method for counter-conditioning. The method aims at severing the connection between a conditioned stimulus and a conditioned response that has developed through classical conditioning. This goal is achieved by arranging confrontations with the conditioned stimulus concurrently with an activity that is incompatible with the conditioned response (fear, in most cases). The activity most widely used to counter fear is relaxation and concentrating on pleasant fantasies. Other behavior also has the same qualities (eating, constructive activities related to the feared object, sexual responses). It is of vital importance for counter-conditioning that the chosen incompatible or irreconcilable response, such as relaxation, is stronger than the anxiety response. For this reason a gradual confrontation is brought about by use of a hierarchy of fear-evoking stimuli.

It is evident that this notion of desensitization relates to the two factor model. After all, it is in this model that the essence of the process of coping with traumatic experiences is expressed in terms of classical conditioning. The persistence of the conditioned response is an expression of inadequate coping with the experiences. The application of desensitization thus implies an attempt to disregard the factors that obstruct the process of coping or, in other words, to achieve (or even force) the extinction that had not occurred up to that moment. Behavior resulting from operant conditioning, considered to be responsible for the obstruction or disruption of extinction, is treated only to the extent that it is directly relevant to the process of desensitization. Sub-assertiveness which can prevent people from experiencing anger, for example, is not treated as such. Only the components that are problematic to desensitization, such as the experiencing and expressing of feelings in therapy, will be dealt with.

Desensitization as training of cognitive skills

The theoretical explanation Wolpe provided for his method was criticized from several angles. For example, his fundamental assumption that a relatively passive process of de-conditioning occurs in desensitization was attacked. In contrast, desensitization may also be interpreted as an active process, aimed at learning a general fear-reducing skill (Goldfried, 1971). This perspective was supported in experimental work (Tirrel & Mount, 1977; Zemore, 1975).

The process of desensitization was conceptualized like this: through certain experiences, an individual has learned to react to situations with anxiety and avoidance behavior. During desensitization, the client learns to become aware of physical feelings of tension, and to react to them with the learned skill of relaxation. In this way he learns to actively deal with fear.

Seligman (1975) emphasized that people do not merely learn responses but that they also become aware of the consequences of stimuli and responses. Similarly, Goldfried argues that cognitive components play an important part in learning. The most important element is not that people learn to relax, but that they adopt the idea that they are able to apply this skill whenever they want. This perspective of the process of desensitization is in line with Seligman's model of learned helplessness. Based on this model, we argue that one of the most important problems in coping disorders is that the client uses internal, general, stable attributions, especially in relation to negative events and external, specific, unstable ones in relation to positive events.

Desensitization enables the client to take control of the situation. He learns that the consequences of traumatic experiences can be coped with through his own skills. The patient's new skills and his 'knowledge' that he is able to control an important part of his life, interfere with the attribution style. Opportunities for more self-confidence and positive self-esteem are thus created.

The approaches mentioned above each indicate a certain aspect of desensitization. In Wolpe's approach the therapist takes on the main responsibility: the responses of the client have to be desensitized. Goldfried's approach accentuates the client's control over his own responses. The therapist takes on the role of teaching the client certain skills and helping him apply them. The preference for a certain form is contingent upon several variables. When the aim is to increase the client's influence upon his own life, a goal that is implicit for most forms of therapy, then the Goldfried approach is preferable. However, there are also situations in which a short-term intervention can be carried out, such as when a client exhibits very dependent behavior. If the initial aim in such cases is to counteract an overwhelmed emotional state, then the Wolpe approach has certain advantages.

11.4 The practice of trauma desensitization

The method described here is a modification of the traditional desensitization procedure. It has been applied successfully in a group of 31 civilian trauma victims as part of a therapy outcome study (Brom et al., 1989). In this study people were treated approximately two years after the sudden loss of a loved person, acts of violence or traffic accidents. We will describe the protocol developed by Defares and Brom (Brom, Kleber & Defares, 1986) that was used in this study. It should be adapted to suit other settings and other groups of trauma victims.

The following elements are important in trauma desensitization:

A. the treatment schedule, i.e. how often and how long will the patient be seen;
B. instruction of the patient at the outset of treatment;
C. exploration and construction of hierarchies;
D. relaxation;
E. desensitization.

A. The treatment schedule

The duration of the treatment, in the case of civilian trauma, such as sudden loss, traffic accidents or acts of violence is approximately 15 sessions of one hour. The decision about treatment length is contingent on the mutual agreement of therapist and client. Keeping the treatment relatively brief in general sustains positive expectation of the patient.
The treatment schedule is as follows:

session 1 – 3	– exploration and creation of hierarchies;
session 4 – 5	– relaxation training using the abbreviated Jacobson method (progressive muscular relaxation); – making an inventory of pleasant images; – desensitization of the item with the lowest value;
as of 6th session	– relaxation training using breathing techniques, in which a biofeedback device developed by Defares (1982) was used; – desensitization according to the hierarchies.

B. Instruction

The initial instruction of patients about the aims and means of a treatment counteracts the anxiety which generally arises at the beginning of therapy. Patients have all kinds of fears about the nature and the effectiveness of therapy. In the first phase it is important to be empathic to these fears and to provide patients with a framework, through permitting them to grasp what kind of treatment is proposed.

Patients with posttraumatic stress disorders in general arrive in a state of high anxiety and lack of control over thoughts and feelings concerning the traumatic incident they experienced. The initial treatment sessions often intensify the intrusion of unbidden images and emotions into consciousness. The instruction should prepare people for this phenomenon and give them a perspective, which helps them to overcome this phase.

The preparation of patients for therapy is an ongoing process, and cognitive instruction is only one part of this. The following is an example of an instruction, with an emphasis on control and on the ability to confront the traumatic event.

> *'In this treatment we want to teach you to achieve control over the tension you experience when you remember unpleasant events. You notice that you tend to become very tense when you think of those events. Consequently, you will often try not to think about them. The aim of this treatment is to think about these events but in a relaxed manner. This relaxation will help you to cope with what has happened.'*

C. Exploration

In the exploration phase of trauma desensitization, stimuli are collected that are directly related to the traumatic event experienced by the client. First, the event which prompted the need for treatment is reconstructed in detail. A process of enquiring into the details reveals the sensitive topics and memories, which the patient avoids. It is only after the entire situation to the smallest detail is apparent that explicit attention is given to the subjective experiences the client had during the event. Finally, an inventory is made also of stimuli that arouse anxiety or are avoided in current everyday life.

In the phase of exploration, the aim is to find 'themes', that is to say stimuli (images, persons, places) with connected feelings. These are separated in such a way that each image represents only one modality of feeling. Once therapist and client agree that enough themes have been found, these are categorized in terms of feelings with which they are connected. In this way separate groups of themes connected to anxiety, anger, guilt and other emotions are created.

The different themes are consequently scored by the client on 100 point

BOX 1

Mrs. W is a 33 year old woman, working in a medium-sized bank in The Netherlands. Three years before her referral she was responsible for the office during an armed robbery. She was threatened with a gun pointed at her chest and it took her a few seconds to realize that this situation was a real one and that her life was in danger. She acted in a cool and organized way, and prevented violence. The robbery was over in 8 minutes. Mrs. W has been suffering from anxiety and from physical symptoms of stress, such as hyperventilation and headaches and was diagnosed as suffering from PTSD.
In the trauma desensitization that was conducted the following hierarchies were used:

Anxiety	Level of Emotion
– Being held with gun on chest	100
– Being held and robber says: 'If you say one word, everyone goes'	90
– Seeing someone with a balaclava (e.g. at a dress-up party)	90
– Thinking about death	85
– Feeling short of breath	80
– Feeling not able to leave (e.g. train)	75
– Seeing violence on TV	70
– Confrontation with attacker at police-station	50
– Sitting behind counter at bank	25
Anger	
– Colleagues and boss saying: 'Think about something else, go back to work'	60
– General practitioner makes fun of her situation	50
– Boss forces her to bend the truth about the robbery in order to protect himself	50
– Lack of understanding and help from friends	30
– Thought: Why always me	25
Guilt	
– Thinking: 'I should have prevented this'	75
– Confrontation with perpetrator: not being sure he was the one	50
Anxiety from previous events	
– Father has a heart attack	25

scales that represent the level of emotion. In this way different hierarchies of anxieties, anger, guilt, sorrow, powerlessness, or other emotions that are deemed relevant, come into being. An example of an actual list of themes is given in box 1. One should be aware that the absolute values of the items in the hierarchies say more about personality characteristics of the person than about the actual tension the items evoke.

In the traditional desensitization procedure one assumes a relationship between different items in a hierarchy. This means that if a strong fear-eliciting stimuli is desensitized, the lower items lose their relevance. In contrast, when creating the hierarchies in this method, no relationship with regard to content between the items is assumed. This has consequences for the ultimate desensitization. When a 'high' item has been desensitized, this does not imply that a lower item has become superfluous. In this sense the adjective 'systematic' has justifiably not been added to the term trauma desensitization.

D. Relaxation

From the fourth session, relaxation exercises using progressive muscular relaxation (Jacobson, 1964) are started. There are many relaxation techniques, and a choice should be made based on the characteristics and preference of the patient. Two specific considerations seem to us important here:

1. to enhance feelings of personal control and to counteract dependence, an active way of relaxation is preferable over more passive techniques, such as guided imagery.
2. in many cases we have observed a relation between posttraumatic stress and hyperventilation. This relation is not surprising, considering the relation between stress and hyperventilation (Janis, Defares & Grossman, 1983; Wientjes, Grossman, Gaillard et al., 1986) Because of this relation an important contribution can be made by the training of breathing techniques.

The most important idea behind the method of breathing regulation is that breathing is a very important determining factor of sympathetic and parasympathetic activity. Moreover, breathing, which has a direct influence on other physiological mechanisms, can voluntarily be controlled by the individual. Introducing a regular breathing pattern is therefore a good method of achieving a condition of physical relaxation.

In order to induce a breathing pattern, the possibility of a 'breathing regulator' exists. This apparatus produces two different sounds, one increasing in frequency, the other decreasing in frequency. The four components of the breathing pattern, that is to say the inhalation, the exhalation and the pauses in between, may each be separately scheduled in terms of duration. The effectiveness of this apparatus was proven in a study on the treatment of hyperventilation (Swart, Grossman & Defares, 1983) and it was also used in the earlier mentioned outcome study on trauma desensitization for civilian trauma victims (chapter 14).

E. Desensitization

The process of desensitization takes place according to the following cycle:

a) relaxation;
b) confrontation with an item;
c) reexperiencing and termination;
d) evoking a pleasant image.

We will elaborate on each step:

a) Relaxation : the client relaxes with or without the help of the therapist. The therapist repeatedly asks the client about the level of tension/ relaxation and adjusts his instructions accordingly. When the client feels he is fully relaxed, the step towards confrontation is made.
b) Confrontation : the client is asked to bring up the image that is connected to a specific item of the hierarchy. It is advisable to work through the items connected to one emotion first, before going to the next hierarchy. One must, however, be aware that sudden changes and breakthroughs occur. When the focus, for example, might be on a 'anxiety item', a sudden shift to anger laden images may occur. The therapist should of course give empathic attention to the intrusive emotions and when possible lead back to the procedure.
   It is important that the client has a clear perception of the image. He can be aided in this by having his attention drawn to details (see also Keane et al., 1985), for example: 'look at... and let the image sink in... look at the color of..., etc.' The experiences show that patients often need several confrontations with the same item, before they see a clear image.
c) Reexperiencing and termination: the client is instructed to relive the feelings that have been evoked. The confrontation is concluded with the following instruction: 'When you feel that you really experienced the situation, you can stop the image by saying aloud 'it is over, I am letting it go'. During this last instruction, pressure should be avoided. The

client works at his own pace and it is he who decides when to cease the imagery by saying the sentence. The way in which the sentence is said can yield important clues for the therapist. When spoken quickly and superficially, it often indicates avoidance rather than experiencing. A lively experience is generally concluded with an emotionally laden expression. Each item, depending on the degree of reliving, will be dealt with two or three times.

d) Evoking a pleasant image: after each confrontation the client is asked to relax and to evoke a pleasant image linked to the item that was just evoked. After themes in which anxiety plays a role, a general image such as a landscape or the beach is evoked. After the confrontation with themes involving guilt, anger or sadness, if they relate to persons with whom a pleasant relationship existed, a memory or image of the pleasant relationship with that person is evoked.

Themes concerning sorrow and grief are not brought up in desensitization. Because of the existentially different value of sorrow, the client is confronted with these themes without relaxation. The sequence in this process becomes confrontation, reexperiencing and evoking a pleasant image.

During the last two sessions, images relating to the future are also evoked. These should be closely connected to the daily life of the client, and they should be a logical continuation of the way in which the client is functioning at that particular point in time.

11.5 Summary

Trauma desensitization is a method in which the client reexperiences a traumatic event he has gone through. The main aim of the method is to restore the feeling of control, which was shattered by the event. Every step in the procedure of trauma desensitization has an aspect of control in it. Dividing the event into themes, teaching relaxation, consciously bringing up traumatic images and also voluntarily stopping the imagery all carry this theme of regaining control within them.

In the theoretical part of this chapter we observe a certain tension. We described the process of coping with traumatic events as a process of classical conditioning. This process can become blocked, when instrumentally conditioned behavior interferes. This combination of two different processes shows once more that posttraumatic disorders are determined not only by the trauma, but also by environmental or personality factors.

This tension brings to mind the well-known question about the determination of the response to trauma. Can trauma alone lead to severe reactions, or is it the premorbid personality that contributes most to the pathology? In earlier chapters we have laid out a model of multiple determination. Behavioral theory clearly supports this model. Although posttraumatic stress disorders must clearly be seen as a response which is directly related to the preceding event, there are other factors that influence the course of the coping process.

From a clinical point of view, however, we should accentuate that it is possible to treat the response to a variety of traumatic events without treating pre-existing personality styles. Trauma desensitization is a treatment modality, that has been used and has proven its applicability and its efficacy.

CHAPTER 12

SHORT-TERM PSYCHODYNAMIC THERAPY

Among the forms of therapy discussed in this book, psychodynamic therapy bears the most resemblance to everyday conversation. The therapist and client talk together. Techniques that are easily recognizable, such as relaxation or homework assignments, are absent. Paradoxically enough, it is also the form of therapy supported by the most extensive theory, and the one that requires the longest period of training in order to practice it consistently.

The theoretical perspective of psychodynamic therapy for posttraumatic stress disorders was developed by the American psychoanalyst Mardi J. Horowitz. His theory on coping with severely distressing events was discussed in chapter 8. It is an elaborate perspective that is clearly linked to psychoanalytical theory, but which, on the other hand, uses concepts that are consistent with recent theory on stress and information processing.

In recent years the cognitive-psychodynamic approach has won much acclaim although definitions may vary. An important shortcoming of the theory is that it does not account for the sharp rift that is observed between the traumatized individual and his surroundings. In an effort to fill this gap, in recent years clinicians have described the response to traumatic events from the perspective of self-psychology. Traumatic events always lead to a certain degree of loss of self-esteem or loss of self-cohesion. Formulating these aspects of trauma refers not so much to the difficulty to integrate the traumatic experience into everyday life, but refers to the damage to the internal organization of the personality. The assessment of this damage is a very important factor in setting out the treatment strategy.

In this chapter we review short-term psychodynamic therapies. These developed as a response to long-term psychoanalytical therapies. Utilizing a restriction of therapy goals as well as a more active attitude of the therapist, a type of short-term therapy has emerged that is based upon psychoanalytical theory.

This chapter has three parts. In the first we discuss theoretical views on coping with traumatic events from the psychodynamic and information processing angle. Then we will go into the theory of trauma as a cause of narcissistic injuries. We conclude the chapter by looking at the practice of brief psychodynamic therapy.

12.1 A general model of coping with trauma

Freud was one of the first to recognize the importance of the concept of trauma in his theories. In Freud's work, this concept appears in a number of different definitions (chapter 2), which are linked to a variety of observations.

Initially trauma was defined as: 'Any experience which calls up distressing affects, such as those of fright, anxiety, shame, or physical pain'. This condition is necessary, but not sufficient for trauma. Freud recognized both the quantitative aspect of the external event (i.e. how severe is the stressor), and the 'preparedness' of the individual in terms of the meaning of the event to the person.

One of Freud's definitions of a traumatic experience is associated with the concept of the 'stimulus barrier'. According to Freud (1920), a traumatic experience causes a breakthrough in the 'stimulus barrier'. The shocking event entails such an over-stimulation of the person that he is defenseless against it. The individual is overwhelmed by the event, in all its affective value and meanings.

Another concept in Freud's early work, is that a trauma occurs when there is a conflict between the ego and an idea that confronts it. This idea is then repressed into the unconscious. While the concept of the stimulus barrier is fairly nonspecific about the events that could possibly cause a posttraumatic reaction, the perspective of a conflict with the ego adds a specificity that can explain why the same event can cause a variety of responses in different people.

Freud made the important observation that, after traumatic experiences, people often reexperienced parts of that event. He referred to this as compulsion repetition.

These ideas have had an enormous influence upon later theory formation on the concept of trauma. They form the basis of Horowitz' theory, in which he links a number of Freud's concepts with concepts from stress theory and from theories on cognitive data processing.

The stress response syndrome

Based upon research on the psychological effects of traumatic experiences, Horowitz (1976) concluded that there is a clear similarity in human reactions to traumatic events. On the basis of these conclusions but also on the grounds of experimental research (Horowitz & Becker, 1971a, 1971b), Horowitz postulated a sequential model of coping, the stress response syndrome. This syndrome occurs to varying degrees in anyone who has lived through a traumatic experience.

As discussed in chapter 2, Horowitz distinguishes between five phases in this syndrome:

1. *Outcry*, the first emotional and almost reflex-like reaction to a shocking event.
2. *Denial*, an intra-psychic mechanism with which the person defends himself against his emotional reaction to the event.
3. *Intrusion*, the almost compulsive reexperiencing of the event in associations, dreams, and fantasies.
4. *Working through*, facing up to the event and the consequences it has upon emotional and cognitive life.
5. *Completion*, returning to the proportions of daily life whereby the significance of the event is integrated into the way the person perceives himself and the world around him.

The most important and specific characteristics of the stress response are phases two and three, namely the mechanisms of intra-psychic denial and emotional insensitivity, and that of reexperiencing, which can occur on the cognitive, emotional, or behavioral level. Although these two phases generally alternate, they also occur simultaneously at different levels.

Three aspects of the coping process will be dealt with separately here. The first is the psychoanalytic perspective of the way people control their conscious experiences. Subsequently, on the basis of the stress and coping paradigm, the emphasis is upon the importance of individual experience and the ways in which people adjust to threatening situations. Finally, based on the theory of data processing, the process of coping is seen as a way of solving an intra-psychic conflict.

The stimulus barrier

As mentioned above, Horowitz elaborated on the theoretical concepts of Freud. The concept of the 'stimulus barrier' is elaborated into a model of the controlling influence exerted upon the process of coping by the intensity of emotional experiences. Schematically, we can picture this system as follows:

Figure 12.1 The stimulus barrier as a feedback mechanism

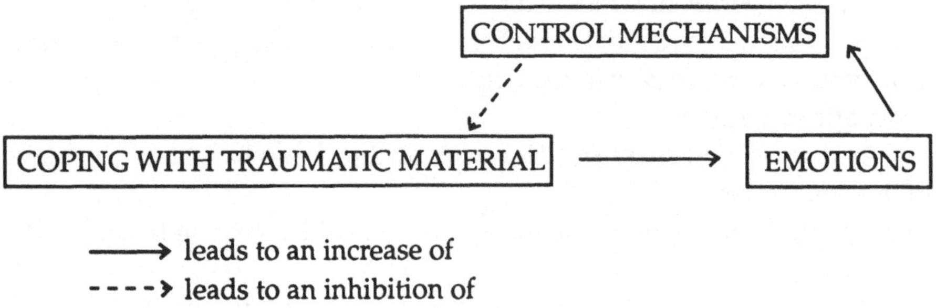

The recollection of an extreme event, which forms a necessary part of the coping process, triggers strong emotions. When these emotions rise above a certain level, control mechanisms will be set in motion which are responsible for temporarily interrupting the coping process. This results in the emotions decreasing in intensity, leading to an increase in the motivation to continue coping.

A more elaborate model of psychoanalytical theory concerning the control of conscious experience is offered in figure 12.2 (Horowitz, 1976). This model contains sources both inside and outside of the individual that offer material for conscious imagery or experience. Control mechanisms may lead to restraint, both of external stimuli (i.e. in this case the increase of the 'stimulus barrier'), and of internal ones (when repression or other defense mechanisms are used).

Emotional experiences have a regulatory function in the process of coping in this model of the psychological phenomena that occur after traumatic experiences. When external or internal stimuli lead to overly intense experiences, intra-psychic measures are taken to decrease the stimulation. Internal stimulation is countered by defense mechanisms that ensure that unconscious material does not reach consciousness. External stimuli are

Figure 12.2 Classic psychoanalytical view of consciousness and control (Horowitz, 1976, p. 84)

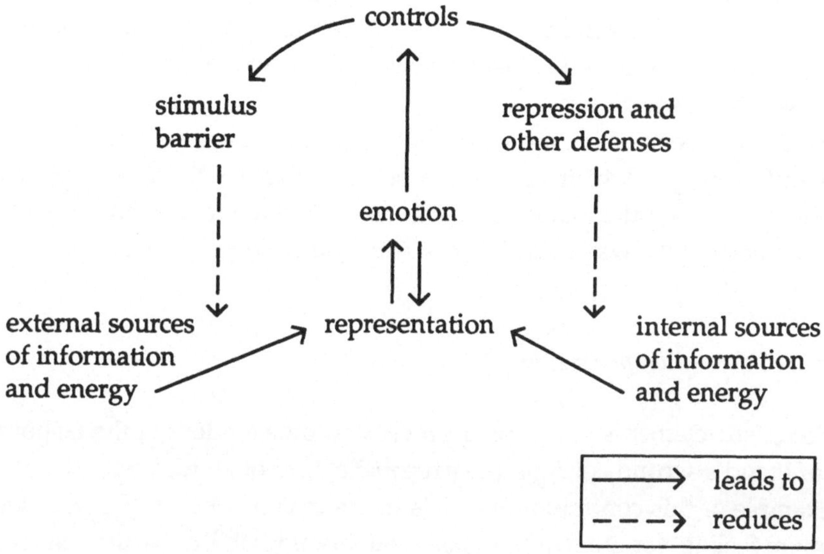

held off when the person has a decreased level of attention and will make himself less sensitive to the outside world. The emotional experiences can be so strong as to cause forms of intra-psychic denial. In such cases, people may even feel that the event did not actually take place.

Stress theory

Lazarus (1981) developed a perspective on the concept of stress, in which the individual interpretation of stimuli takes on a central position. He identifies several processes of interpreting stressful situations. We discussed these mechanisms in chapter 8. In 'primary appraisal' the individual decides whether or not a threat exists. When a threat is perceived, 'secondary appraisal' takes place, which focuses on the capacities a person feels he has for dealing with the threat. These ideas can be combined with the above-mentioned psychoanalytic view concerning the controlling of consciousness. Threatening situations lead to anxiety, but it is not the situation alone that determines the fear but also the cognitive appraisal of the danger. This 'reflective fear' (Janis, 1969) leads to three possible modes of adjustment.

One mode of reaction is alertness, in which a person is continuously on guard for threatening stimuli. A high degree of alertness can easily lead to startle reactions, one of the symptoms that are considered part of the process of reexperiencing. The need for reassurance is another form of reaction. In extreme cases, this reassurance may involve elements of denial and insensitivity. A third mode of adjustment consists of a compromise between the latter two types of responses.

This stress perspective indicates that there is no direct link between an event and the way in which a particular person will react. Both the appraisal of the situation, and the totality of reactions evoked in the individual by the event, determine the way in which a person will attempt to adjust.

Cognitive information processing

Characteristic elements of a coping process become evident in the elaboration of Freud's second concept, the irreconcilability of an idea with the ego. An example of this confrontation exists in the case of victims of crime. The common feeling (or illusion) of personal invulnerability, which enables people to function freely, is incompatible with the realization, that such an act of violence can happen to a person any moment. Horowitz elaborates on this idea using theories on the integration of information found in the work of Piaget (chapter 8). Each traumatic experience confronts the person with information that was not present up to that time.

How will someone deal with such new information? How does a person alter his thinking and functioning in a way that leaves room for the reality of the event, but still allows for continuity in daily life? The answer may be found in a closer examination of the concept of compulsive repetition. Freud identified two elements of compulsive repetition. One of these is instinctive in nature, and led him to the postulation of the notion of the death wish; the other is a need for gaining retroactive control over the situation. Horowitz especially emphasizes this latter element. He views the process of reexperiencing in the context of the ego function, i.e. reconciling the consequences of the event with existing cognitive schemata. Both cognitive and emotional/behavioral reexperiencing can be viewed as resulting from a need to process new information to the point of completion (chapter 8).

The information-processing model as applied to the stress response syndrome operates on the following assumptions:

1. Storing new information in active memory implies that it will continually return to conscious experience.

2. Only the completion of the process of integration will make this information disappear from active memory.

The interaction between all sources of information is schematically represented in figure 12.3

Figure 12.3 The integration of information as a way of ending compulsive repetition (Horowitz, 1976, p. 97)

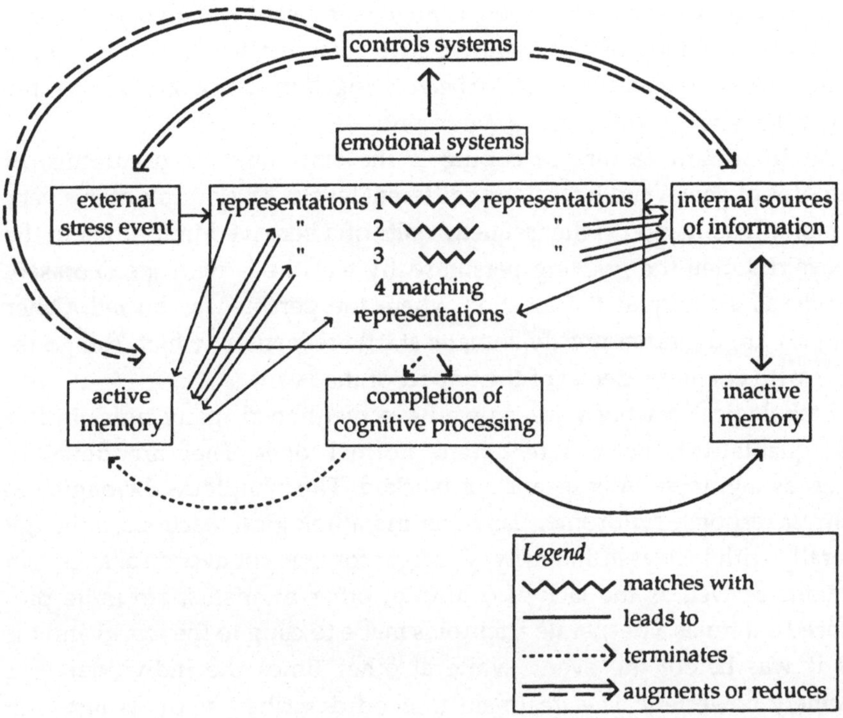

Summing up, an extreme life event results in information that is in conflict with ideas the individual already holds. This leads to a process in which the representation of the conflicting material is repeated until integration becomes possible. This is indicated in figure 12.3 as the comparison between the new information and the representations stemming from internal sources. During the process, the conscious experiencing and repression of the conflict alternate because of the fluctuation of the emotional experience. Intense emotional experiences lead to repression and less intense experiences make room for the conscious representation of the conflict. By changing the meaning a person assigns to the event, the intensity of the emotions decreases with time, and a greater accessibility is created for conscious experience. Normally, in a stress response syndrome, a number of phases follow the event:

1. complete suppression of the representations;
2. alternate reexperiencing and repression, which result in a gradual process of coping with the experience;
3. completion of the process of coping, which ends the cycle of reexperiencing and repression.

Completion is achieved when there is no longer a conflict between the new information and the existing cognitive schemata: in other words, when the existing ideas and conceptions have been changed in such a manner that the old and the new information are congruent.

An important feature of coping is the simultaneous occurrence of symptoms of denial and intrusion at different levels. An example of this very common phenomenon is the frequent visits of a bereaved individual to the graveyard. Often the grieving person really feels he is with the deceased, when he is actually at the location where the person was buried. Other examples are the often mostly somatic startle responses, which can go together with cognitive denial of the impact of the event.

Pathological reactions are, as we have mentioned many times in this book, qualitatively not different from 'normal' ones. They are, however, longer lasting, more intense, and/or blocked. The symptoms, belonging to the stress response syndrome, also occur in pathological reactions, although generally with increased intensity. The fight for emotional control at certain moments as well as the lack of control at other moments, are more pronounced. At times a desperate attempt is made to cling to the world and the way it was before the event, while at other times the individual feels extremely powerless. The treatment method described below is aimed at reducing both polar reactions.

12.2 Traumatization and self-injuries

The approach presented above successfully accounts for many of the symptoms that characterize the process of coping. Some symptoms, however, are not explained in a satisfactory manner. How do we account for the feelings of estrangement, which are so common in victims? What about the breach with their environment? And how do we deal with the lowered self-esteem? All of these symptoms revolve around the survivor's relationship with his social world.

In an attempt to deal with these issues, trauma theory has broadened in recent years to include ideas raised by self-psychology. These ideas complement the cognitive or information processing approaches to posttraumatic stress disorders. We will outline the concept of trauma as a narcissistic injury briefly but should note that the terminology in this field is not very clear. Concepts, such as narcissistic injury, self-injury and narcissistic regression are used interchangeably.

Kohut (1977) developed a structural approach to psychological development, which has proven to be of great value in understanding narcissistic personality disorders, and in planning and conducting their treatment. Moreover, this approach has also proven to be of great importance in the treatment of disorders of the self, which express themselves in various clinical syndromes.

Some authors have adopted the theoretical principles of the treatnent of disorders of the self and applied them to the field of trauma. Catherall (1989) makes a distinction between the primary and the secondary trauma. He conceptualizes the primary trauma within ego-psychology and maintains that there are specific symptoms, that are connected to the primary trauma. The symptoms related to the overwhelming affective nature of the traumatic event and the symptoms related to the individual's efforts to control the internal sequelae of the event are all expressions of the difficulties in ego-functioning caused by the event.

The secondary trauma is conceptualized as the breakdown of the relationship between the victim and his environment. In previous chapters we mentioned acute disruption as one of the essential features of traumatic experiences. Many survivors experience a sharp breach with their social world. In the literature on the Holocaust this is expressed as belonging to 'the other world' or 'the other planet'. Survivors feel that no communication is possible with people who 'were not there'. Symptoms of the secondary trauma are social withdrawal, lack of pleasure in previously pleasurable activities, feeling of alienation, lowered self-esteem and interpersonal difficulties.

There is a conceptual difference between treating traumatized individuals with narcissistic personalities and treating narcissistic injuries in traumatized clients. Horowitz (1986) relates to the former, when he writes about the application of brief dynamic therapy to people with narcissistic personalities. Marmar and Freeman (1988), however, discuss the possibility to apply brief dynamic psychotherapy in the treatment of narcissistic regression following traumatic events. Goldberg (1973) went even so far as to propose a separate diagnostic category for acute narcissistic injuries.

What is a self-injury or a narcissistic injury? Narcissism is not only present in personality disorders, but is a basic human asset. The self is seen as a 'cohesive and enduring psychic configuration', i.e.

> *'the basis for our sense of being an independent center of initiative and perception, integrated with our central ambition and ideals and with our experience that our body and mind form a unit in space and a continuum in time' (Kohut, 1977).*

Like other aspects of human personality, there is a continuum of possible disturbances in the self, ranging from a completely integrated, independent and stable personality to severely disturbed, fragmented personality disorders.

The origins of narcissism lie in childhood. The utterly chaotic world of the infant is slowly structured through the early relationship with the parental figures. The infant experiences the world partly through the parents, while the parents function as so called 'self-objects'. In a healthy development, the self-objects slowly transform into aspects of the self. The admiring responses of the parents, their capacity for mirroring and their availability for merger with the infant are necessary conditions for a healthy self-development.

Disorders in the development of the self exist in varying degrees and constitute the narcissistic vulnerability of the person. The narcissistically vulnerable individual searches, although not consciously, for relationships which complement the unfulfilled needs, in order to compensate for the feelings of fragmentation and weakness. A common way to compensate for these feelings is an inflated self-concept. When we deal with narcissistic vulnerability and less with a narcissistic personality disorder, the latter defense of the grandiose self may not be prominent at all. The depressed and inhibited states may be more striking.

Traumatic experiences always lead to a certain degree of damage in self-esteem, narcissistic injury, or loss of self-cohesion. In more extreme traumatic experiences, such as the concentration camp or severe combat, there is a massive threat to personality organization. Fragmentation of personality under the influence of massive trauma can be seen in dissociative states and reactive psychoses. In these instances the central personality organization is, at least temporarily, shattered or split up.

All people are vulnerable to narcissistic injuries and react by a decrease in self-esteem, a decrease in functioning and social withdrawal. The severity of the reaction is dependent upon the stability and cohesiveness of the pre-trauma personality. Ulman and Brothers (1988) maintain that the meaning of trauma is 'unconsciously determined by the shattering and faulty restoration of 'central organizing fantasies' of archaic narcissism' (p. 176). The less the archaic narcissistic fantasies are transformed, the more vulnerable the person is to an event which threatens to shatter these central organizing fantasies of self.

The approach to trauma described above has developed in the past years in different forms (McCann & Pearlman, 1990; Parson, 1986; Van der Kolk, 1987). On the basis of their various theoretical elaborations the above authors outline and recommend different treatment strategies. The next section is devoted solely to the contribution of this line of thought to short-term psychodynamic therapy.

12.3 Short-term psychodynamic therapy

Both theory and practice of psychoanalytic treatment have undergone substantial development. Freud's early therapeutic treatment was neither long-term nor aimed at insight. Alexander and French (1946) outlined the development of the psychoanalytical form of treatment by Freud and in doing so also indicated the reasons for the increasing length of time for the duration of therapies. They identified five phases in this development:

1. *Cathartic hypnosis*. In this phase, Freud discovered that hysterical symptoms originate from emotional disturbances in the past, and that events that trigger these disturbances can be completely repressed from consciousness. He used authoritarian hypnotic techniques to bring repressed memories back to the surface, and thereby evoke emotional discharges. He concluded, however, that these discharges did not result in a permanent reduction of symptoms.
2. *Suggestions during consciousness*. Freud discovered that not everyone was susceptible to his techniques of hypnosis, and replaced them with direct suggestions. He told his patients that there had undoubtedly been a certain traumatic experience they had thus far forgotten, but would be able to remember. Freud did not have much success with this technique either; he encountered resistance from his patients.
3. *Free association*. In an effort to avoid this resistance from patients, Freud developed a new strategy, namely that of free association. He asked the patient to say anything that came to mind. In this way he hoped to retrieve repressed traumatic experiences. During this phase, the goal was still to bring the repressed material to the surface so that the emotions could be reexperienced. The emotional discharge that would take place in this gradual way was intended to yield lasting results.

4. *Transference.* Freud's most important discovery, according to many, was the concept of transference. The reaction a patient exhibited towards Freud was, in his view, a repetition of earlier contacts with individuals that were important to the patient. The reexperiencing of these relationships, aimed towards the therapist, and the interpretation of the transference became very important principles in the therapy.
5. *Emotional re-education.* In this last phase, the emphasis was not so much solely on events within the therapy sessions, but on the importance of the influence of therapy upon becoming involved in new experiences in daily life. During this period it became clear to Freud that the therapy could be viewed as a catalyst. The rediscovery of memories in this light is not a goal in itself, but serves as an indication of the degree to which the patient is able to permit memories to take place and thereby allowing for new experiences, so that a 'corrective emotional experience' is possible.

The passivity of the therapist was one of the main factors that lengthened the psychoanalytical process (Alexander & French, 1946; Malan, 1963). A second factor that lengthened the process was that the interpretation of the transference, which delved further and further into the life of the patient. Ultimately, relating transference to early childhood experiences and relations became the ideal.

As far back as the twenties, there were already analysts who questioned the constantly increasing duration of the therapy. The most important among them was Ferenczi (Ferenczi & Rank, 1926) who particularly criticized the passivity of the therapist. In later writings, however, he became milder in his criticisms and pointed out the limitations of short-term therapy.

The idea of brief psychodynamic therapy faded into the background until the publication in 1946 of the work of Alexander and French. They argued that many analysts of their time became locked into a phase that Freud had ignored: many analysts still strove for cathartic experiences. In contrast to this, Alexander and French did not consider the accessibility of memories which had been repressed up to that point as the goal of a process, but as its result. They viewed the therapeutic process as one of learning, in which the ego learns new ways in which to handle certain feelings. They argued that not only is a trauma 'dealt with' during therapy, but that a person also learns to cope with traumatic experiences.

As of the fifties, the interest in short-term therapy on a psychodynamic basis increased. In England, a group of therapists, under the direction of Balint, began using this form of therapy. In 1963, Malan described the work and ideas of this group. At the same time, the American Sifneos developed his ideas and method of treatment.

The practice of short-term psychodynamic therapy

There are three questions that, according to Malan (1963, 1976), each form of short-term therapy should deal with. These are the selection of clients, the technique, and the results. What changes can be achieved in which clients with what techniques?

With regard to therapy of a psychodynamic nature, two extreme standpoints may be distinguished. From a conservative viewpoint, short-term therapy was only suitable for clients with mild disturbances that had recently surfaced. The techniques had to be superficial, and interpretation of the transference was to be avoided. In terms of results, there should not be high expectations as to the results; the therapy would have a palliative effect, that is to say that the intra-psychic conflicts and, consequently, the life of the patient, would be made less painful. As the conflicts would not be resolved, lasting results could not be expected. The radical new perspective argued that real change could indeed be achieved in clients with relatively serious and chronic disturbances. The therapist was to engage in active interpretation, and all elements of a complete psychoanalysis would be utilized.

Sifneos and Malan both initially adhered to this radical perspective. In 1976, however, Malan concluded that the sharp contradiction between conservative and radical had disappeared. Spectacular results can be achieved with some patients, while with others, the results are only palliative. Alexander explained some dramatic positive results in terms of the existence of a 'benign trauma' that dramatically removes pathology, just like a bad experience may create pathology. This fairly unspecific opinion has in the past years been replaced by a more specific concept of the process and effects of brief psychotherapy (Gustafson, 1981).

Selection of clients

Sifneos (1972) utilized very stringent selection criteria for his anxiety provoking therapy. It may well be that he could only maintain these because he worked with a group of highly educated, 'high class' clients at Harvard. Some of his criteria were:

1. above average intelligence;
2. ability to interact well with the intaker and to express feelings verbally;
3. a specific primary complaint.

Malan (1976) is, however, more radical in his views and emphasizes the possibility of selecting, at an early stage, a focus as the most important selection criterion. In addition, the client should have sufficient abilities to achieve psychodynamic insights, and his motivation is another important variable. The gravity of the pathology is not of as decisive importance as is a criterion of selection. Malan's criteria are now more generally accepted (Rogawski, 1982).

During the intake phase for this kind of therapy, immediate attention is paid to the possibility of formulating a central conflict in the client's problems. A central conflict is a psychodynamic concept through which the client's complaints can be explained. In such a theme, earlier and recent traumatic experiences, the early family situation, and important behavior patterns serve as basic elements. The central conflict is the basis for the choice of a focus. Cassee, Petrie and Wolf (1984) view the focus as 'a clearly defined area upon which the therapist will concentrate during therapy'. The focus can fully coincide with the central conflict but it is also possible that it will only cover a part of that conflict.

The potential of the client in terms of his reaction to interpretations is also assessed during the intake phase. Using trial interpretations, the intaker will determine the extent to which a good therapeutic attitude can be expected of the client within a brief period of time. The question whether the client can handle interpretations constructively is of great importance.

Finally, an appraisal will have to be made of the client's motivation. Motivation for psychotherapy is a complex issue. On the one hand it is often used by therapists to explain bad results of a treatment and on the other hand it is not easy to assess in a consistent way. Elements of motivation include the patient's willingness to see the symptoms as being psychological, the skills to change oneself and to take an active part in the therapy, realistic expectations of the outcome, and the willingness to invest time and money in the treatment.

Technique

In short-term therapy, the therapist adopts an active attitude. Generally there are three phases present in short-term therapy. During the first phase, the focus is defined and a working alliance is established. Sifneos (1981) maintains the therapeutical alliance by interpreting resistance, ambivalences, and negative transference symptoms at an early stage.

As the relationship between the therapist and the patient becomes more secure, the therapist's attention shifts more towards leading the patient in the direction of the focus. Not only the content of conversations, but also the interpretations of feelings of transference are related to the focus. Malan stresses the importance of establishing connections between transference feelings and the relationship of the client to his parents. He refers to this as the 'transference parent link' and considers it to be the ultimate effective ingredient in short-term therapy. During the middle phase of therapy, especially in the twelve session therapy of Mann (1973), the initial positive expectation is fading and patients are bringing up their ambivalence towards brief therapy.

The third phase of therapy is that of the termination. The ending of therapy and the separation from the therapist evokes many emotions. There is often a reexperiencing of the central conflict (Cassee at al., 1984). The atmosphere should make this reexperiencing a corrective emotional experience, while it is of great importance that negative transference phenomena are discussed thoroughly at this stage.

Mann (1981) has taken the issue of time as a central point of attention. The clear duration of the treatment brings about a rapid confrontation between the infantile and often narcissistic perception of time and the real time in connection with the reality of the treatment. Through this paradoxical confrontation of the time issue, an impetus is given to the transference and to the working alliance.

Results

The goals of short-term dynamic therapy differ according to author. All authors emphasize the solution of a neurotic conflict that is at the basis of the client's problems. Malan (1979) is concerned with the clarification of the three components of the conflict triangle. By clarifying the mechanisms of defense and the anxiety, the hidden impulse will ultimately be brought to consciousness. Malan's method of treatment may bring about a fairly radical personality change. This has been proven over time. But in the majority of cases one should expect symptomatic relief and the management of a stressful conflict in an adaptive new way (Gustafson, 1981).

Sifneos (1972) adds a specific element to the goal of the therapy. In line with Alexander and French, he speaks of a learning process that occurs during therapy. The client learns how to solve new emotional problems.

Selection, technique and results in Horowitz' work

Horowitz has specialized in the treatment of a specific group of clients, namely those who suffer from disturbances in coping with traumatic experiences. In principle, his work corresponds with that of other authors in the area of short-term dynamic psychotherapy. A difference, however, is that Horowitz addresses himself explicitly to alleviating the current disturbances in the coping process and, in fact, does not aim to achieve personality change. Although personality growth is an ideal goal in his treatment method, he is only moderately optimistic about its feasibility.

Horowitz and Kaltreider (1980a) compare the working relationship between therapist and client to a path. On one side of the path there is the danger of social involvement, in which sufficient openness for therapeutic work is lacking. On the other side there is the danger of becoming involved in strong transference reactions, which can not be resolved in brief therapy. These dangers are possibly even greater during the treatment of posttraumatic stress disorders. The strong feelings of powerlessness and abandonment following most traumatic events, inevitably crop up in therapy. The client's appeal for advice and support has a real dimension, but may just as easily be a test to determine if the therapist would abandon him.

The working relationship between therapist and client is viewed by Horowitz in a specific context. He regards it as a relatively secure and agreed upon model of roles and rules to which the client and therapist will adhere. Transference reactions will, when necessary, be contrasted with the working relationship. The client's perception of the therapist can actively be put beside the perception of the therapist as a therapeutic ally. Moreover, the client's self-image in transference can be compared to his self-image in the working relationship. Horowitz hereby prevents the escalation of transference reactions, and offers his clients the opportunity of reality testing.

This use of the working relationship also corresponds with Horowitz' view of the learning process. He believes the client learns to deal with experiences, not only through insight based on interpretations, but also through identification with the therapist. During a given period of therapy, the client often reverses the therapeutic roles, and renders the therapist powerless, just as he was rendered powerless by the traumatic event. The way in which the therapist deals with this can help the client to confront repressed feelings. In this way he 'copies the therapist's coping manoeuvres'

(Horowitz & Kaltreider, 1980b, p. 74.)

During the second phase of therapy, after a secure relationship has developed, Horowitz focuses on changing the client's control mechanisms. We previously described that clients go through periods of denial and reexperiencing after traumatic experiences. According to psychodynamic theory, these periods are related to an excess of cognitive control (denial) or to a lack of sufficient control (intrusive reexperiencing).

When the client experiences a lack of control, the therapist will aim at strengthening it. The main way in which control can be reinforced is by making so-called 're-constructive interpretations' that help the client sort out the facts, perceive relationships and distinguish reality from fantasy. In addition, the client is assisted in directing his attention, sometimes to the traumatic experience, and sometimes to other things. The client's ability to determine the degree of exposure to the material in this manner strengthens the feeling of control. In short, the therapist combats the client's lack of control through an active, supportive attitude.

Excessive cognitive control is expressed by the denial of parts of the traumatic experience, and the avoidance of activities that stir up memories of it. The individual's activities are aimed at keeping the emotions at bay. In this case, the therapist's interventions will be aimed at decreasing control. Emotional experiences are encouraged by identifying defense mechanisms, and stimulating detailed descriptions and association. The therapist may adopt a more confronting attitude.

Both attitudes of the therapist, the supportive and the confronting, should not be exclusive attitudes. We described earlier how people, at a given moment, show both symptoms of intrusive reexperiencing and symptoms of denial on different levels. In terms of treatment, this means that one cannot speak of absolute phases of either reexperiencing or denial, but rather of the relative predominance of symptoms of one phase or the other. The therapist will therefore have to respond very flexibly to the constantly changing balance of the poles of reexperiencing and denial. The therapist's compass will be aimed at making a coping process possible, that is to say that the traumatic experience and all its accompanying emotions will have to be confronted, without the balance swinging over to any one side. If the client is capable of this, the therapist will guide the treatment towards the process of working through the meanings associated with the experience.

The third and last phase concerns the termination of treatment. This will generally involve a reexperiencing of the traumatic event and accompanying symptoms. It is only logical that termination of a short treatment stirs up feelings of abandonment, which are so very common also during and after traumatic events. The time required for the coping process is much longer than the duration of the treatment and the treatment can only be an impetus for coping. The loss of the relationship with the therapist will therefore provoke anxiety and anger. Provided that the termination of the treatment is referred to at an early stage, it can create the possibility for the client to deal with the loss gradually, and in an increasingly active manner. Interpretation of the connection between experiences in connection related to the end of treatment and the traumatic event, can considerably reinforce therapeutic effects.

In this chapter we discussed the method of treatment based upon psychodynamic theory with an emphasis on cognitive information processing. The main aim of the therapist is to facilitate ego-growth, so that the traumatic material can be integrated into the pre-existing cognitive schemata. When we adopt the view of trauma as creating narcissistic injuries and as a threat to self-cohesion, this affects our treatment method.

The first aim in this type of therapy is to provide the patient with a self-object relationship and to respect the regressive tendencies of the patient. Accurate empathy is the key concept in the treatment of the traumatized individual. Being able to feel the alienation and the rift between society and the patient is a crucial condition for helping the individual regain self-cohesion. Initial acceptance of the patient's need for idealizing and being idealized will help build the relationship.

Empathic failures play an important role in treatment. Empathic failures may cause another regression, which becomes evident in a lack of confidence or in feelings of anger or rage on the part of the client. If properly handled, the empathic relation is re-established. Proper handling means that the therapist is willing to take full responsibility for empathic failure, when it occurs, without losing sight of the specific sensitivities of the client. The continuous movement to and from the therapist, if in a condition of 'optimal frustration', helps build a more cohesive and stable self, that relies more on inner sources of strength and independence.

Reconstructing the event remains an important aspect of the treatment, but with narcissistically vulnerable patients this is a particularly critical activity. Untimely interpretations very easily disrupt the patient's confidence in the therapist (Marmar & Freeman, 1988). After establishing and consolidating the relationship, the therapist's reconstruction of the traumatic event should gradually help the patient tolerate the bad self-images and the accompanying emotions.

The goal of the treatment along the lines of self-psychology is the re-establishment of the ties with the outside world. One of the dangers, which is perceptible within circles of traumatized people, is that the traumatization provides a pseudo-identity. From the traumatic event onwards people consider themselves to be 'bereaved parents', 'Vietnam veterans' or 'victims'. The acceptance of this pseudo-identity may lead to a consolidation of the inability to connect to the outside world. Group treatment of victims of a specific kind of traumatic event is particularly vulnerable to this danger.

12.4 Concluding remarks

This chapter described short-term psychodynamic therapy for the treatment of disorders in the process of coping with traumatic events from two angles. First, we described the well elaborated perspective of coping as an adaptive mechanism, in which cognitive features stand out. Secondly, we went into the more recent development, which sees trauma as narcissistic injury. Finally, we tried to integrate this background with the ideas about brief psychodynamic therapy. The result is a rich theoretical view.

The extent to which trauma theory and the background of focal psychodynamic therapy are truly reconcilable remains uncertain. There is considerable ambiguity concerning the manner in which a focus should be chosen. Viewed in terms of the trauma issue and Horowitz' view of it, there is a tendency to concentrate on a conflict in relation to the traumatic event. However, if we adopt Malan's view, it is possible that a totally different focus would emerge. In any case, it is striking that Horowitz mentions little about this choice of focus, and therefore also about this problem.

A second question raised by Horowitz' theory is the extent to which his system may be too closed. One could also obtain a similar impression from Horowitz' practical work. Although the therapist's compass may be emotions, the steering wheel constructed by Horowitz is cognitive processing. This implies the danger that the client's control is strengthened at a highly cognitive level, while it remains to be seen whether this control is, or will be, supported by an appropriate emotional working through.

Finally, we should comment on the time perspective in the treatment of trauma victims. The advantages of brief therapy are very appealing. The time perspective in brief therapy, which is so important in the work of Mann, is certainly important in the work with traumatized patients. Very often the short duration causes the reexperiencing of the feelings of abandonment that were experienced so sharply during the traumatic event. However, the decision regarding the duration of therapy should be taken with caution.

The depth of the narcissistic injury is, in our opinion, an important factor in this decision. If the narcissistic injury or regression takes a relatively important place in the clinical picture, then brief therapy is less appropriate. The more threat there is to the central factor of self-cohesion, the higher the risk is for splits in the personality, and thus the less appropriate is a short, focus oriented therapy.

CHAPTER 13

HYPNOTHERAPY

Most people are familiar with a condition in which attention is so concentrated that the environment seems to fade and whatever is concentrated upon becomes so important one feels a part of it. We all know the feeling of 'getting into' a movie or of identifying with an athlete to such an extent that one involuntarily mimics the movements of the other person. These experiences are referred to as trance experiences. Since the beginning of psychotherapy, people have attempted to use trance experiences in the treatment of psychogenic complaints.

Hypnotherapy is practiced from different perspectives and with different goals. Some therapists call themselves hypnoanalysts and base their work on psychoanalytic theory. Hypnosis, however, can also be found in behavioral therapy. In the past years the use of hypnotherapy for posttraumatic disorders has grown enormously.

In this chapter we review the use of hypnosis in the treatment of disorders in the process of coping with traumatic experiences. After a brief historical outline, we discuss with some theoretical visions on trauma and hypnosis. Finally, we describe some practical issues of hypnotherapeutic treatment.

13.1 History

It is said that hypnosis is as old as the world itself. In rituals used by primitive tribes, hypnotic phenomena used to take place and continue to do so (Kroger & Fezler, 1976). The first use of hypnosis as a method of treatment in our society is attributed to Mesmer. Mesmer was interested in the healing effect of magnets as well as in 'animal magnetism', to which he attributed the positive effects of touching those parts of the body that were unwell. In the 19th century, Mesmer's work was followed by a number of physicians. Braid, the English physician, was the first to link the word hypnosis to the condition which, in his opinion, was caused by fatigue of the eye muscles after prolonged eye fixation. Braid later began to assume that hypnosis was based not so much on a physiological effect, but on a reaction to the suggestions of the hypnotist. Through Braid, a first step was made toward the recognition of hypnosis by the medical world of that day. In later years the work of the neurologist Charcot generated an even wider acceptance of hypnosis.

Freud also used hypnosis in his treatments, but discontinued this practice for a number of reasons. Chertok (1984) describes an incident during one of Freud's sessions with a female client in 1892. At the end of a session utilizing hypnosis, the woman threw her arms around Freud's neck. Freud explained this behavior by assuming the presence of a third person in the (unconscious) experience of the patient, the transference figure. According to Freud, it was not simply a reaction to him, but also a reaction to another person projected on him. In order to observe these projections as clearly as possible, the therapist should abstain from actual suggestions. One of the reasons for discontinuing the use of hypnosis was that phenomena of projection during hypnosis could not, according to Freud, be controlled to such an extent that they could be analyzed and explained.

According to many, this decision made by Freud was the reason that psychoanalysts chose not to use hypnosis as a method until the fifties. Some psychiatrists and psychologists, however, did continue to apply and research hypnosis. As a result, both the method of hypnosis and the theory relating to the nature of hypnotic experiences developed. One of the major changes in the application of hypnosis was the method of hypnotizing. Until the end of World War II, hypnosis was induced in an authoritarian fashion (Fromm, 1984). The patient was told, often even ordered, what he should do and feel. Later on, a more 'democratic' way of hypnotizing was developed in which more space was left for the individual needs and preferences of the patient.

13.2 What is hypnosis?

We see a person sitting in a chair. He looks very relaxed, his eyes are closed, his breathing is very deep and regular. The person sitting opposite him has his eyes wide open and watches the other carefully. He observes the movements in the other's body, mentions these and links them to a suggestion of increasing relaxation and heightened attention to everything happening 'within'. Especially the small movements occurring in the hands are pointed out. The suggestion is made that the small movements could be the beginning of an upward movement of the arm that occurs more or less automatically. After a while, the arm does move upward in small, jerky movements, while the person remains in a fully relaxed state. It is said that with the raising of his arm, the person will go increasingly deeper into a trance and will more readily allow things to happen that are beneficial for him.

There are many opinions on the nature of the above phenomenon, hypnotic induction via the method of hand levitation, as well as on all other hypnotic phenomena. These opinions generally express a theoretical orientation that entails more than just a view of hypnosis.

From the psychoanalytic perspective, links are made between the hypnotic relationship (between therapist and client) and important relationships in the life of the subject. Chertok (1984) views the hypnotic condition as a reliving of the early symbiotic mother-child relationship. Fromm (1980) includes the developmental phases of the psychological apparatus and speaks of adaptive regression.

Other theoretical perspectives provide different views of the phenomenon of hypnosis. From the point of view of learning theory Hull (1933), for example, argues that hypnosis can be viewed as a habit. From the perspective of 'role theory', Sarbin and Coe (1972) use terms such as role play to indicate what goes on between hypnotist and subject.

In addition to conceptions stemming from these wider theoretical backgrounds, there are very specific descriptions of hypnosis that rely less on existing theory. Erickson (1952) describes hypnosis as a condition of intensified awareness and receptivity as well as an increased responsiveness to an idea or a set of ideas. Hilgard (1970), however, emphasizes the dissociation between different cognitive systems, which occurs during hypnosis.

It is apparent from the above that no unambiguous definition of the phenomenon of hypnosis is currently available. For our current purposes it will suffice to point out certain important conditions for the occurrence of hypnotic phenomena. First of all, there must be a good relationship between hypnotist and subject. The subject must be prepared and able to accept suggestions made by the hypnotist. This readiness is often described as motivation. The possibility to accept suggestions depends, among other things, on the degree to which a person can be hypnotized. Hypnotizability, like hypnosis itself, is a concept that eludes precise definition. During hypnosis, a narrowing of consciousness and a selective alertness occurs. While the environment appears to fade for the subject, the object of concentration is experienced more intensely. Finally, one is more open to sensations from within during hypnosis. Fantasy and imagery can more easily be drawn on.

13.3 Theory

Psychotherapy is conducted on the basis of a theory of human functioning. Currently there is strong trend to deny the importance of theory and to work from an eclectic perspective, but this only obscures the issue. Psychotherapy without theory is unthinkable. A therapist, implicitly or explicitly, decides on what to focus and what to neglect in his relation with the client. These decisions, even if not conscious, are based on theoretical presuppositions.

Because hypnosis itself does not derive from theory, but rather is a collection of phenomena, we present several theoretical views on trauma and connect them to matching views on hypnosis. Three theoretical approaches to trauma and hypnosis are discussed in this chapter. In all three of them we relate to the following case description.

Client B, a 37-year-old male, is employed as a prison guard in The Netherlands. Eighteen months before commencement of the therapy, B was involved in large scale fighting inside the prison. The fighting broke out when a prisoner was removed as the result of drug use and after repeated refusal to hand in his cigarette. Two guards tried to take him to an isolation cell by force, but the prisoner resisted and a general fight broke out. B held on to the prisoner with all his might but felt he was hurled to and fro beyond control. B was hit several times and from a certain point on he experienced the event as if in a mist: 'as though I was watching myself and everything that happened on TV'. The fight got out of hand, and the riot police had to come in to make an end to the battle. B does not remember the events after the arrival of the riot police until he was sitting in the guard's unit.

B was the fourth child in a family of seven children. The atmosphere within the family was cool and B learned very early in his life to take care of himself. Although he was not particularly tall for his age, he developed in a street fighter at an early age. He always felt pity for the underdog, and when smaller kids were in trouble he used to take their side. When he would really get angry and let himself go, he was very strong and hardly ever tasted defeat.

B was not a motivated pupil and left school at an early age to go to work. He worked in the construction industry until an economic recession sent many laborers home. Then for some years he worked as a barman in different bars until he was unemployed. He found a job as prison guard after many job applications. Prison guards have a low status in The Netherlands.

B was tense since the incident, but continued working as usual, although he noticed that he tended to avoid situations, which could lead to confrontations with prisoners. He felt guilty about the event and avoided his superiors. Eighteen months after the incident he was assigned to take a troublemaking prisoner to an isolation cell. When he opened the door of the cell, the prisoner stood in front of him in a threatening posture. A split second B felt the impulse to hit, but then, without hesitation B slammed the door, walked to the director and offered his resignation. The director referred him to the social worker and soon he was referred for psychotherapy.

The main complaints of B were:

- *tension and irritability in his work situation. He feared a repetition of the events;*
- *anxiety, especially in the work situation but also outside of it;*
- *reexperiencing the event. The client experiences this as viewing 'a movie' of the events, which he cannot stop; afterwards he feels anxious;*
- *physical tension, specifically muscle aches and shortness of breath;*
- *sudden feelings of panic;*
- *decreased self-confidence.*

Learning theory and the 'hypno-behavioral model'

In chapter 11 we reviewed several ways to conceptualize posttraumatic reactions within the framework of learning theory. Without repeating the basic principles, we will look upon the reactions of B from the behavioral angle. In particular, we will try to understand his reactions with the help of the two factor model of Mowrer. The initial traumatic situation triggered a process of conditioning; the anxiety, tension and intrusive reexperiencing are (classically) conditioned reactions to aspects of the original situation. When approaching the ward, where it happened, B experienced physiological reactions, such as accelerated respiration and heart rate. B also felt intense anger toward the prisoners and in particular toward the prisoner who had caused the upheaval. The anger surfaced very often in connection with thoughts about the whole situation.

Avoidance was the main way in which B tried to cope with his fears. He avoided situations that reminded him of the original situation and tried to work as little as possible in direct contact with prisoners. These (instrumentally) conditioned behaviors were strengthened by their power to prevent the conditioned physiological and emotional reactions. His deeply embedded need to be in control of his emotions served as a background variable, strengthening his avoidance behavior.

The connection between learning theory and hypnosis was made by Kroger and Fezler (1976). They view hypnosis merely as a technique. They refer to their approach as a 'hypno-behavioral model'. Hypnosis, according to Kroger and Fezler, is a collection of phenomena related to suggestion and trance. They do not arrive at a definition of hypnosis but provide us with an extensive summary of hypnotic phenomena and of methods used to achieve them. These phenomena, such as the increased accessibility of imaginary experiences, are viewed as the most important characteristics of hypnosis.

Kroger and Fezler describe behavioral therapy, both in terms of techniques and theoretical background. They integrate hypnosis and behavioral therapy by combining their techniques. They do not provide a theoretical background of hypnotic phenomena and techniques and this makes the postulation of a hypno-behavioral model appear somewhat exaggerated. Behavioral therapy is the theoretical background, but in practice much attention is paid to imaginary experiences. The vivid mental imagery, which clients achieve by using hypnosis is valuable and facilitates procedures such as sensitization and desensitization.

Treatment of B within the hypno-behavioral model would mean that the aim is to bring about a confrontation with the stimuli, which B tends to avoid. Hypnosis can help here, by creating a safe situation and an easy access to imagery.

Dissociation

The work of Pierre Janet was mentioned in chapter 2. This philosopher/psychiatrist was the first who developed systematic theory about the connection between traumatic events and psychopathology. His work was almost forgotten. The revival of the interest in his work in recent years is related to the revival of hypnosis. Janet relied heavily on the use of hypnosis in the treatment of disorders after traumatic experiences.

Janet's basic idea (Van der Kolk, Brown & Van der Hart, 1989; Van der Hart & Friedman, 1989) is that trauma results from a failure to take effective action against a potential threat. His view resembles the cognitive-psychodynamic vision, described in chapter 12, to a certain degree. The helplessness, resulting from vehement emotions, interferes with storage of the experience in memory. The crucial factor determining whether memory storage will or will not take place, is the fit between the personal significance of the event and the existing cognitive schemes.

Janet's view parts from psychodynamic thinking when it comes to the interpretation of the posttraumatic phenomena, such as the tendency to deny or avoid thoughts and feelings about the traumatic events. Psychodynamic theory deals with repression and other defense mechanisms, suggesting a clear relation between repressed material and integral personality functioning. Janet proposes dissociation as explanatory concept. Memories of traumatic events may be dissociated from conscious awareness. This explains, for example, the partial amnesia, which frequently occurs in clients with posttraumatic stress disorders. Dissociated material may return in conversion reactions, or in other maladaptive ways. Through dissociation vertical splits in personality are brought about and new cores of consciousness come into being.

For Hilgard (1970), dissociation refers to the suppression of aspects of experiences in relation to momentary consciousness and conscious control. It is related to cognitive as well as to neuromuscular aspects. Dissociation is a broad concept that applies to phenomena from 'normal' life and to psychopathological ones. Listening to someone and simultaneously formulating an answer would be an example of dissociation. At the extreme end of the continuum of phenomena of dissociation lies the multiple personality (Young, 1987), where a total split exists between different so called ego-identities.

Going back to B, it is clear that his reactions fit the definitions of dissociation. Even during the fight he felt disconnected from reality and watched himself as an uninvolved outsider. His amnesia of elements of the events is another clear sign of dissociation. The loss of control during the fight and the near loss of control of his anger nine months later serve very well as the underlying conflict, leading to the dissociative states.

Hypnosis also implies the dissociation of physical and emotional sensations. The level of involvement in the imaginary activities is high enough for the person to temporarily dispense with the process of testing his experiences in terms of reality. It is thus possible that people under hypnosis can accept two contradictory suggestions ('trance logic').

During hypnotic induction, the therapist will make suggestions that produce the feeling that certain processes happen 'automatically'. A vivid example of dissociation is found in the technique of hand levitation. This technique, which may be used as a method of induction for good hypnotic subjects, requires that the subject pays attention to one or both hands. The involuntary movements that occur in the hands are labeled by the therapist as activities of the hand without the conscious involvement of the subject. This enhances the experience of dissociation of the hand from the main stream of consciousness.

The perspective of hypnosis as 'controlled' or 'guided' dissociation makes the use of hypnotherapy for posttraumatic disorders attractive (Kingsbury, 1988; Spiegel & Cardena, 1990).

The psychodynamic approach and hypnosis

The psychodynamic approach to trauma was described in chapter 12. Two perspectives make up the psychodynamic approach. The information processing side on the one hand, claims that the experience has to be integrated into existing cognitive schemes and this is done through a search for meaning. Self-psychology, on the other hand, conceptualizes trauma as a narcissistic regression.

Erica Fromm has studied hypnosis and its applications since the mid-'50s. Her work is characterized by a broad vision inspired by psychoanalysis and experimental research. She suggests that hypnosis is an altered state of consciousness and that the relationship between hypnotist and client is of principal importance. She views the ability to be hypnotized partially as a personality trait, and partially as a skill that can be developed.

Fromm (1980, 1984) illustrates the process of hypnosis with the aid of three parameters:

1. *Primary versus secondary process*
   Human functioning is viewed as a result of changes in 'cognitive structures', such as perception, cognition, attention, memory and feeling. The organization of these structures determines what an individual is attuned to. Children are focused on the immediate gratification of needs without knowing the limitations and consequences imposed by external

reality. Consequently, children think more in the form of fantasies (primary process) than in the form of logical, structured images (secondary process). As the demands of reality are integrated, conceptual thinking becomes more prominent. Both poles occur in adults and continuously interact. Primary process functioning plays an important role during hypnosis.

2. *Ego activity, ego passivity and ego receptivity*
   In psychoanalytic literature, a distinction is made between the ego, the id and the superego. The ego is related to functioning through the application of the reality principle, the id is related to functioning through the application of the lust principle and the superego is related to the norms one feels should be upheld by oneself and others. Ego activity implies that the ego undertakes action either to maintain control over the environment and the internal impulses, or to diminish the consequences of a partial loss of control. Ego passivity occurs when the individual surrenders to the pressure of the other psychological structures and feels helpless in the face of the demands that are made.
   A third orientation of the ego is receptivity. Here, the ego is not strongly directed toward reality or specific purposes, but is more open to unconscious and preconscious material. At one moment a person may attempt to manipulate the environment for certain goals (ego active), but the next moment the person may let things happen (ego receptive). Fromm and Hurt (1980) view ego receptivity as a state that facilitates insight.
3. *Reorganization of cognitive structures*
   The third parameter important in hypnosis is the reorganization of certain cognitive structures, whereby a selection of information takes place through the senses. During hypnotic induction, one of the processes that occurs is a narrowing of attention for stimuli.

The approach of Fromm corresponds very well to the main features of psychodynamic theory on trauma. Hypnosis can be helpful, both as a means to uncover conflicts and work them through, and as a means to allow regression in order to compensate for narcissistic injuries.

The treatment of B with this approach would undoubtedly focus on the issues of control over emotions and of aggressive impulses. B's life history revealed a need for rigid defenses on the basis of early neglect. The incident, which triggered the development of a posttraumatic stress disorder, taxed one of the weak spots of B, i.e. the fine dividing line between loss of control/aggressive impulses and his strong need to be in control. His work as prison guard in general stressed this point over and over, but until the incident he had been able to use his aggressive impulses when the situation required it. The incident brought him in a state of uncontrolled rage, and at the same time it cut him off from his social environment and he felt alone and abandoned.

13.4 Treatment

We return to the treatment of B.

> *The first three sessions were devoted to the exploration of his present situation, his past and the traumatic incident. In the fourth session the therapist proposed to use hypnosis. B felt both curious and anxious, and said he was afraid of losing control. The therapist agreed with him to take a gradual approach of relaxation, in order for B to get used to letting go at his own pace. In the first inductions his own control was a main element of the suggestions (e.g. choose a safe place of your own, where you can allow yourself to let go).*
>
> *The client proved to be an excellent hypnotic subject, easily capable of vivid visualization and fantasy. By focusing on bodily sensations he was capable of achieving deep trance levels. The therapist decided, therefore, to use open evocative suggestions. The client was asked, for example, to walk in his imagination up a hill. The therapist suggested that he might be surprised at what he might see when he would reach the top of the hill. The client became very tense when he was approaching the top. While in a medium level of trance, images that emerged were those of mountain scenery and events that symbolized the control of the client over the situation. He perceived himself to be very big in comparison with the landscape. Standing at the top of the mountain he felt an enormous relief and could relax while looking at the scenery. He was impressed by this experience.*
>
> *The next session B came in saying he felt much better. He felt very much in control and had the impression that he was cured. He asked the therapist to terminate the treatment. The therapist suggested to use hypnosis to look back at what had happened in the prison and see whether his feeling of control would hold there. B agreed and easily went into a rather deep trance. At the suggestion that he could go back to the prison he became very tense. He reported walking in the prison in the direction of the ward. The next moment he was in the middle of the fight, that had taken place nine months before, and he became visibly anxious and showed symptoms of*

> *hyperventilation. The therapist suggested that he was participating in a film (an image B had previously used to describe his intrusive memories) and that he could step off the screen and look at the film instead of participate. B readily accepted the suggestion and calmed down. He saw the fight, but was surprised that he did not recognize the prisoners. They seemed to be faceless.*
>
> *B was now very confident and in subsequent sessions he proved capable of transforming his experiences under hypnosis. Gradually working towards reexperiencing, the event itself, including the relevant emotions, was alternated with comforting visualizations. This instilled confidence in the client in terms of his ability to control the moments during which he was overwhelmed. In dealing with the events, partially at home during self-hypnosis, it became clear that the memory itself had undergone change. While only the fighting itself was reexperienced during the first part of therapy, after a few more sessions B appeared to be able to see what went on around it, only to notice that many prisoners actually were not involved in the fighting, and were not even interested in it. Previously, he thought that all prisoners had participated in the fight.*

The three theoretical views presented in this chapter may be applied to this treatment. The analysis of Brett and Ostroff (1985) describes the role of imagery in the different theoretical models of posttraumatic stress disorder and clarifies the process of cure in the case of B. According to conditioning theory, a gradual confrontation with the traumatic event took place and this desensitized the fears and tensions that were connected with the event. From this point of view the treatment could have been desensitization or exposure. However, the imagery in this treatment gave an extra dimension to this treatment, which was very helpful.

The dissociation perspective looks differently at what happened. Like many people suffering from posttraumatic stress disorders, B proved to be easily hypnotizable (Spiegel & Cardena, 1990). This may indicate his pre-existing tendency to use dissociation. The gradual and guided integration of the dissociated emotional experiences in the therapy are the most important features of this treatment (Van der Hart, Brown & Van der Kolk, 1989).

From a psychodynamic point of view, B suffered from a disorder in the coping process on the basis of a strong narcissistic vulnerability. The emotional neglect in his childhood has made B into a man with a strong need to be in control and to find an acceptable way to express strong aggressive impulses. The event brought him in contact with primary rage, which in turn necessitated the use of dissociation as defense mechanism.

Hypnotherapy provides the possibility to offer a relationship between therapist and client which fulfills primary needs. The caring and soothing experience of relaxation seems very suitable for people with narcissistic features. The exclusive attention, the authority attributed to the therapist and the regressive feature of trance experiences contribute to this suitability.

13.5 Practical aspects of hypnotherapy

Considerable changes have occurred over the past decades in the practice of hypnotherapy. It used to be the therapist who controlled events during a session, whereas nowadays much more freedom is given to the client to determine the goals he wishes to attain and which issues should be addressed. Gilligan (1981) mentions three approaches to hypnosis:

- the authoritarian approach, in which the therapist uses his suggestive or authoritative power to induce trance;
- the standardized approach, based on receptivity to hypnosis as a personality trait, uses standard procedures to achieve the trance state;
- the utilization approach, which is ascribed to Milton H. Erickson (1952), and is based on interactive aspects in the hypnotic relationship. It is not so much the authority of the therapist, but primarily the qualities of the subject that are important. Instead of using standard methods, the therapist addresses the question of how he can facilitate trance experiences in the client. This approach recognizes the joint responsibility and effort of therapist and client as the basis of hypnosis, and it is this perspective that formed the basis for the hypnotherapeutic treatment described and proposed in the last section of this chapter.

Hypnosis can be helpful in several forms of psychotherapy. Our emphasis here is on the conditioning model. This treatment aims at confronting the client with the total reality of the traumatic event, in order to achieve a decrease in the conditioned reactions to it. The treatment of predispositions for the occurrence of disturbances in the coping process, is not included in the goals.

In essence it is possible to treat posttraumatic stress disorders caused by 'simple' traumatization in treatments of around 15 sessions. In terms of the actual treatment, the following aspects are to be discussed by the therapists:

1. Exploration

About three sessions are devoted to the exploration of the problem. First, the event is reconstructed in detail. Sensitive issues and avoidances especially become clear during the description of details. Second, once the event has become clear to the therapist, additional attention is given to the experience of the event and to the exploration of aspects which are most difficult for the client. Toward the end of the exploration, a list is made of issues that are important to the client. Images of incidents and of persons are differentiated in such a way that each image represents only one modality of feeling.

2. Instructions for treatment

Therapists may use the following instruction to express their view of the disorder:

> *'If, after a lengthy period of time, an event still bothers you, this means you have not fully dealt with it. Thoughts of this event and the feelings that go with it have been partially suppressed, not experienced. It was all too much and you preferred not to feel the pain. The aim of this treatment is to no longer suppress these feelings, but gradually attempt to experience what it all means to you. In this way you will be able to cope with such an event.'*

A number of common-sense ideas exist about hypnosis, some of which increase existing anxieties and obstruct its helpful usage in psychotherapy. An adequate explanation of hypnosis is important and can be used to increase motivation. An attitude of positive expectation on the client's part can be achieved through suggestions such as:

> *'Hypnosis is a way of letting your feelings roam free while staying in control of them.'*

3. Induction

A good way to prepare induction is exploring the ways in which the client usually relaxes. Does he see images that help him to relax and do these images evoke physical (kinesthetic) sensations or is the client more sensitive to auditory stimuli? A technique of induction is chosen in accordance with the nature and preference of the client. The client is asked to direct his full attention to a certain sensation. Subsequently, the suggestion is given that, as concentration increases, so does relaxation. Involuntary movements made by the client are incorporated into the suggestions and are considered signs of increasing relaxation. The blinking of the eyelids, for example, may be suggested to signify that the eyelids are getting heavier and beginning to close. An inventory of induction procedures is presented by Udolf (1981).

Therapists should be aware of the easy association between relaxation and the traumatic experience. Authoritative suggestions may be interpreted as bearing likeness to the behavior of a perpetrator. Prolonged relaxation, on the other hand, may bring back the feelings of loneliness and abandonment, felt immediately after a traumatic event. The following example may illustrate this:

P was a 38-year-old South American refugee, who spent some years in a concentration camp. He was tortured frequently. As one of the ways to survive the torturing with the least damage he used to make his body as limp as possible. This bodily relaxation, which he achieved with the utmost self-control, helped him endure brutal physical violence. When the therapist started to use relaxation, P was cooperative, but he became noticeably quiet and suspicious. Only after he raised the issue of quitting therapy, he told the therapist about the association with the technique he used during torture.

4. Confrontation

During this phase, the issues that have previously been noted as traumatic stimuli, are brought up and the client is asked to occupy himself with such an issue, either in words, images or in other sensory experiences. The aim is to sustain the confrontation with the subject long enough for the emotional response to gradually decrease. It should be kept in mind that a short confrontation may trigger a sensitization. By ending the confrontation with an emotional response prematurely, avoidance behavior is rewarded. Thus it is important to sustain the confrontation long enough to enable the emotional response to decrease. It is possible that the decrease takes place in stages and not in a linear fashion.

The avoidance of confrontation assumes different forms. The formation of images can be inadequate and ambiguous, or no images emerge at all. This may be illustrated by using the case of the aforementioned client B.

During one of the imaginary confrontations with the fight in prison, B was able to see and feel right up to the moment during which he had actually panicked. Suddenly, however, the image disappeared and the client remained calm. The therapist noted the calmness and only by bringing this up detected the avoidance manoeuvre.

After a few sessions, the client again went through the same moments. The image, however, did not disappear, and the incident advanced. Upon inquiry, however, it became clear that the faces of the prisoners in question, all of whom he had known for a long time, had become unrecognizable to him.

These examples show the importance of the involvement of the therapist in the experiences of the client. The therapist who is not continuously aware of what the client goes through, and of the quality of his experiences, runs the risk of passing over the most intensive stimuli.

13.6 Final remarks

The term hypnotherapy refers to the collective methods of treatment in which hypnosis is used. There is no clear definition of hypnosis. It is clear, however, that a specific sequence of interactions between therapist and subject is essential for hypnosis (Sarbin & Coe, 1972).

The hypnotic relationship creates an 'experienced reality' that lies outside 'everyday' or 'normal' reality. Dissociation can be seen as an interaction between therapist and subject in order to create a distance between experiences in therapy on the one hand, and the experiences in the 'real world' on the other. Suggestions such as 'your hand will automatically want to move upwards', define the situation as an unreal one. In this situation old norms and meanings can be abandoned and one is more open to new or hitherto hidden experiences.

Coping with trauma entails a process of attributing a meaning to the event(s). Disorders in coping are characterized by either a lack of meaning to what has happened or by a meaning that makes life into a frightful existence. The hypnotic relationship enables people to find new meanings. Meanings that were too charged can be reviewed without becoming too threatening. The structures of experience can thus be adjusted in such a way that these experiences also become bearable in the 'normal' world.

Once more we stress here that no theoretical perspective has been developed in hypnotherapy on human functioning and its related problems. The expectations of therapists and clients about hypnosis differ strongly. Failure to recognize these differences will result in ambiguity. The goals, set in terms of a psychodynamic approach, may differ considerably from those aimed at in behavioral therapy. We present this difficulty to urge caution. Theory is indispensable if one wishes to arrive at a framework within which consistent therapeutical work is possible.

CHAPTER 14

THE EFFECTS OF BRIEF PSYCHOTHERAPY

The term posttraumatic stress disorder (see chapter 2) describes psychological symptoms resulting from serious life events that substantially hinder normal functioning. In preceding chapters we have analyzed the disorder as a stagnation in the psychological process of coping with extreme stress. Psychologists, psychiatrists, and other members of the mental health profession are increasingly becoming aware of the prevalence of the disorder (Davidson & Smith, 1990).

Although an extensive literature exists concerning the adjustment to traumatic events, little research has been conducted on the effectiveness of specific psychotherapeutic methods for the treatment of posttraumatic stress disorders. Lindy, Green, Grace et al. (1983) conducted an evaluation of thirty therapies, but they did not use a control condition and the majority of treatments were interrupted prematurely. Horowitz, Marmar, Weiss et al. (1984) studied the outcome and the process of the treatment of 52 bereaved patients. These patients all received time-limited dynamic psychotherapy. In both studies patients started treatment within the first year after the traumatic event.

In this chapter we describe a controlled study of the effectiveness of the three psychotherapeutic methods which were discussed in the chapters 11, 12 and 13. The method of the study will be outlined and the results of the therapies will be dealt with. A separate paragraph is devoted to the issue of indications and contra-indications for the three therapeutic methods (a detailed discussion of this study is found in Brom, Kleber & Defares, 1986; results have also been reported in Brom, Kleber & Defares, 1989).

Numerous studies have been conducted exploring the effectiveness of psychotherapy in general (Smith, Glass & Miller, 1980). Because the results did not reveal many differences in effectiveness among different methods, it is considered necessary to specify both the treatments as well as the research objectives. This call for specification as well as the dearth of similar research on the treatment of posttraumatic stress disorders make the evaluation of the effectiveness of psychotherapy after traumatic experiences a useful undertaking.

We hypothesized that therapies specifically aimed at posttraumatic stress disorders were effective in reducing symptoms related to the disorder. We also assumed that the therapies would not influence personality traits. These assumptions are tested with the use of a waiting-list control condition. In addition we conducted explorative research into which indicators exist for improvement during psychotherapy.

14.1 Method

Sample

The sample consisted of 112 people who were diagnosed as suffering from posttraumatic stress disorders according to DSM-III-R, with the condition that not more than five years had elapsed since the traumatic event. Of the participants, 79% were women, and 21% were men, with ages ranging from 18 to 73 (M = 42.0, SD = 14.3). The majority of participants were married (59%); only 2% were divorced. The widows and widowers (24%) almost all applied for help because of the death of their partner. The remaining 15% were single. The mean level of education was 3-4 years of high school. The scale for professional status (Jager & Mok, 1971) indicated that the group could be considered lower middle class. Fifty-one percent of the participants at the time of the interview occupied a job outside of their household.

The sample consisted of 19 persons who had experienced a violent crime, 4 who were involved in a traffic accident, and 83 individuals who had lost a loved one as a result of murder/suicide (17), traffic accidents (17), acute illness (31), or chronic illness (18). The person who was being mourned was in nearly all cases a member of the immediate family, and some cases involved the death of more than one member of a family. Six patients experienced an event that did not fall under one of these categories.

The level of psychological distress at the pretest indicated that most of the patients were in crisis at the time of their application. In comparison with a group of phobic patients (Arindell & Ettema, 1981), our group proved to have statistically significant higher scores on somatic symptoms, state anxiety, hostility, and psychoneuroticism but a lower score on phobic symptoms. The scores on the Impact of Event Scale were considerably higher than those reported by Horowitz, Wilner, and Alvarez (1979). The scores on the personality questionnaires were compared with those of patients of general psychiatric outpatient wards. Our patients proved socially more skillful, less rigid, and they had a higher self-esteem than this group. Although these differences point in the direction of less pathology, the scores of our patients on trait anxiety and trait anger point in the opposite direction. Although these differences are statistically significant, their significance in absolute terms seems limited.

In conclusion we can state that the general picture of our sample is lower middle class, neurotic, and with crisis-like symptoms.

Procedure

Two admission interviews were conducted by one of the authors (D.B.). The first interview was a general assessment of the patient, and the traumatic event was discussed. The second interview was conducted to observe the reaction to the first in order to make sure patients could stand a confronting therapy. In this interview the course of the patient's life history was discussed. The 112 selected persons were randomly assigned to one of three therapy conditions: psychodynamic treatment (N = 29, 2 therapists), hypnotherapy (N = 29, 2 therapists), trauma desensitization (N = 31, 3 therapists), and the waiting-list group (N = 23).

The treatment was carried out by therapists who were trained and who had more than 10 years of experience in the specific method they conducted. Each therapist conducted the one form of therapy that he or she preferred outside of the research setting. In order to assure adherence to the procedures supervisory sessions by senior advisors were held (P.B. Defares, trauma desensitization; O. van der Hart, hypnotherapy; M.J. Horowitz, psychodynamic therapy). The mean length of treatment was 15.0 sessions for trauma desensitization (SD = 2.9), 14.4 sessions for hypnotherapy (SD = 1.4) and 18.8 sessions for psychodynamic therapy (SD = 2.6).

Measurements were taken before, immediately after, and 3 months after treatment; the waiting-list group was measured before and after a waiting period of 4 months. The patients in the waiting-list condition received treatment outside of the research setting.

Measures

We focus in this chapter on the data from the standardized questionnaires, disregarding the physiological and behavioral tests which were administered. The fields covered by the questionnaires were general symptoms, symptoms of the coping process, and personality.

General symptoms were assessed by means of the Dutch version of the Symptom Checklist-90 (SCL-90), which was validated by Arindell and Ettema (1981), who obtained a dimensional structure of the following five subscales: (a) Social inadequacy (inadequacy in interpersonal relationships, negative frame of mind, sense of inferiority); (b) Somatization (physical complaints); (c) Agoraphobia; (d) Hostility (symptoms of an aggressive nature); (e) Psychoneuroticism (the sum score of the 90 items). Cronbach's alpha-coefficients ranged between .74 and .96. In addition, a sixth dimension was used that was based on the findings from the literature in the area of complaints (Kleber et al., 1986) that develop after traumatic events. This dimension, which we refer to as trauma symptoms, consists of 27 items that have bearing upon fears, negative emotional experiences, tensions, concentration and memory disturbances, lack of interest in the external world, and sleep disturbances. On this dimension no validation data are available at present.

The State-Trait Anxiety Inventory and the State-Trait Anger Inventory were translated and validated for The Netherlands (Van der Ploeg, 1980; Van der Ploeg, Defares & Spielberger, 1981; Spielberger, Gorsuch & Lushene, 1970). The reliability coefficients of the four scores range between .85 and .91.

The symptoms of coping with extreme stress were assessed by the administration of the Impact of Event Scale (Horowitz et al., 1979), which was translated and validated by Brom and Kleber (1985). The two subscales of intrusion and avoidance present in the original scale were reaffirmed in our study with minor changes, and reliability scores were .72 and .66 (in a second sample, they were .81 and .78) The external validity of the scale is reported elsewhere (Brom et al., 1986).

Characteristics of the personality were assessed by the Dutch Personality Questionnaire (Luteijn, Starren & Van Dijk, 1975), a thoroughly investigated and widely used instrument, comprising the following subscales: (a) Inadequacy (feeling anxious and depressed); (b) Social inadequacy (incompetence in contact with others); (c) Rigidity; (d) Discontentment (suspicious of and hostile towards others); (e) Conceit (satisfied with oneself and not wanting to have anything to do with others and their problems); (f) Dominance (desire to be superior to others); (g) Self-esteem.

In addition, the Introversion-extraversion scale of the Amsterdam Biographical Questionnaire, another well-documented instrument developed and tested by Wilde (1970), was used. Finally, the scale for internal versus external control developed and tested by Andriessen (1972) was used. Cronbach's alpha-coefficients for these measures ranged between .80 and .89, with one exception of .69.

14.2 Results

Dropouts

We regard as dropouts all patients with whom the decision to start a treatment was agreed on, but who discontinued this treatment against the advice of the therapist. A total of 12 participants discontinued treatment in this manner. This amounts to 11% of the total number of accepted patients. This is low, in comparison with the available figures about premature withdrawal from therapy. Garfield (1978) reported dropout percentages of between 30 and 65%. Our low percentage evidently has to do with the brief duration of the therapies, the specific nature of the complaints, and the well defined structure that was offered. The 12 dropouts were evenly distributed over the treatment conditions and received a mean number of six sessions. The dropouts did not differ significantly from the remaining participants in terms of sociodemographic background, symptoms, or personality characteristics (univariate F values ranged from .0 to 3.0, mean F value = .4, p values ranged from .96 to .08).

Analysis

The data were analyzed in four steps: 1. a multivariate analysis of variance (MANOVA) in order to minimize familywise error rates; 2. the comparison in one test between the effects of the treatment conditions and the eventual changes in the same variables in the waiting-list condition; 3. the analysis of the effect of the treatment in comparison with the waiting-list group, controlling for the differences in initial scores. These steps constitute a rigorous testing of the effectiveness of the treatments. And finally, 4. the analysis of various indicators for the success of the treatments.

Before entering into the analyses, we offer some comments on the interdependency of the variables. Using so many variables undoubtedly creates a certain degree of interdependency. We conducted a principal component factor analysis with varimax rotation on the pretreatment variables, which yielded six factors, explaining 72.6% of the variance. Criterion for assigning a variable to a factor was a factor loading of over .60 on one factor and under .30 on the other factors. The most important factor includes most of the variables concerning symptoms. Intrusion and avoidance, however, were found to occupy separate factors. Emotional state was another factor, including state anxiety and state anger. The remaining two factors consisted of personality measures. Because this analysis upheld the general partition we maintained (general symptoms, symptoms of coping and personality), and for the sake of clarity and replicability, we decided to present the variables as they were measured. One should be aware, however, that symptoms can be considered under the heading of neurotic symptoms and emotional state, that intrusion and avoidance are relatively independent, and that the personality measures comprise variables related to social functioning and related to self-esteem.

Multivariate analysis of variance

The first analysis we present here is a multivariate analysis of variance (MANOVA). This analysis serves as a check whether we still have statistically significant results if we enter the variables in one overall test. We entered the most important variables in this test and found a Box-M value (21, 183) of 92.3 ($p = .14$), indicating that our data comply with the specifications for a multivariate analysis of variance. Hotellings' test yielded a value of .52 with a p value of .05. These data do not provide insight but protect against interpretation of one-way analyses of variance, like the ones we will use, without a sound basis. As a result of this analysis our study meets methodological requirements.

The main findings are reported in Tables 1, 2 and 3 of this chapter. In these tables the simple t tests between pretest, posttest and follow-up measurements can be seen, as well as the t tests between the raw difference scores of a treatment condition and the control group and the results of the t tests with the use of the residual gain scores, which we will describe here.

An important methodological drawback of difference scores, composed of raw scores, concerns the high correlations between these difference scores and the scores on the pretest. These impede the interpretation of the results. A solution for this problem is the calculation of so-called residual gain scores. These residual gain scores, based upon the actual differences between pre- and posttesting and on mean group improvement, give an indication of the 'actual' improvement, without being related to the scores on the pretest. The formula with which we have calculated these scores was derived from Meltzoff and Kornreich (1970) and reads as follows:

$$\text{Residual gain} = \frac{Z_1 - R_{01} * Z_0}{1 - R_{01}}$$

whereby Z_1 = transformed posttest score, Z_0 = transformed pretest score, and R_{01} = correlation between raw scores of pre- and post testings. For this calculation the raw scores are converted into Z-scores.

Because the residual gain scores are not very illuminating measures, an analysis of variance was also conducted as a check on these results. In this case the posttest score was introduced as the dependent variable, the condition (i.e. in each case one therapy as opposed to the waiting-list condition) as independent variable, and the pretest score as covariate. This procedure resembles the residual gain score analysis and yielded results that were so similar that only the former results are presented here.

Direct tests between the therapies without the control condition regarding the effect scores on symptoms or personality measures did yield only one significant difference in univariate analyses of variance, indicating that we should consider the therapies equally effective (mean F value (df = 2) = .9 with a mean p value of .40). The comparison on some variables between one therapy condition and the control group yielded significant results, and that between another therapy condition and the control group yielded nonsignificant results. This creates difficulties in interpretation. In the next description we refer to the comparisons between each therapy condition and the control group as they appear in Tables 1, 2 and 3.

Symptoms of coping

The symptoms of intrusion and avoidance, which are central elements of the process of coping, lessened considerably in the treatment groups, but not so in the control group. At the post measurement, the effects of the psychodynamic therapy seem fewest; but these effects appear to continue, so that at follow-up measurement they match those of other therapies. Trauma desensitization and hypnotherapy have a stronger influence on the symptoms of intrusion, and psychodynamic therapy has more influence on the symptoms of avoidance.

Table 14.1 Symptoms of Coping

Therapeutic technique	Intrusion M	Intrusion SD	Avoidance M	Avoidance SD	Total M	Total SD
Trauma desensitization						
Pretest	24.1	5.3	18.9	9.0	47.4	12.0
Posttest	14.7[abc]	9.8	10.7[abc]	8.9	28.0[abc]	19.5
Follow-up	16.0[ab]	9.5	12.3[a]	10.4	31.3[ab]	21.1
Hypnotherapy						
Pretest	25.7	4.6	20.5	8.0	50.8	11.7
Posttest	17.1[abc]	10.5	12.9[ac]	10.7	33.7[ac]	22.9
Follow-up	15.7[ab]	10.9	12.5[a]	10.4	31.7[ab]	22.0
Psychodynamic therapy						
Pretest	23.8	7.1	18.0	10.2	46.3	13.5
Posttest	18.4[a]	8.3	12.0[ac]	8.6	32.7[ac]	16.5
Follow-up	15.0[a]	8.8	9.7[ab]	7.6	27.0[ab]	17.0
Waiting list						
Pretest	24.2	5.8	22.3	6.9	51.1	14.1
Posttest	22.3	6.4	20.5	8.7	46.5	15.2

a *p* value of the *t* test on the difference with the pretest is less than or equal to .05.
b *p* value of *t* test on the pretest-posttest or pre-follow-up differences between treatment and control is less than or equal to .05.
c *p* value of the *t* test on the residual gain scores between treatment and control is less than or equal to .05. (Residual gain scores are only calculated for the pretest-posttest differences.)

General symptoms

In the treatment conditions there is a general drop of the scores of almost all the symptom dimensions. The control group shows slight but not statistically significant improvement. The direct confrontation of the treatments and control group, however, reduces the number of significant outcomes. The use of residual gain scores further reduces the number of statistical results. The psychodynamic therapy seems to withstand the comparison best. Although some differences between the treatment methods can be observed in the data we presented, these differences seem only to have

Table 14.2 General Symptoms

Therapeutic technique	Social inadequacy		Somatization		Agoraphobia		Hostility
	M	SD	M	SD	M	SD	M
Trauma desensitization							
Pretest	19.2	8.0	38.1	12.2	14.7	7.3	7.6
Posttest	15.8[a]	7.1	30.2[a]	13.1	11.4[a]	6.0	7.0
Follow-up	16.4[a]	9.1	31.7[a]	13.0	11.3	6.7	7.3
Hypnotherapy							
Pretest	20.0	7.1	41.4	14.0	16.8	7.4	10.7
Posttest	17.2	9.0	33.3[a]	18.8	13.3[a]	6.5	8.3[ab]
Follow-up	15.9[a]	7.7	30.8	17.2	13.0	7.1	7.1[ab]
Psychodynamic therapy							
Pretest	20.2	7.1	41.6	12.7	16.9	8.5	10.5
Posttest	15.0[ab]	6.2	29.7[a]	12.4	11.7[a]	6.5	7.4[ab]
Follow-up	13.4[a]	6.3	26.6[a]	13.2	10.4[ab]	5.6	6.1[ab]
Waiting List							
Pretest	17.1	8.1	38.4	11.0	13.6	5.6	8.0
Posttest	16.8	8.1	33.8	11.5	11.6	5.5	8.0

a p value of the t test on the difference with the pretest is less than or equal to .05.
b p value of the t test on the pretest–posttest or the pre-follow-up differences between treatment and control less than or equal to .05.
c p value of the t test on the residual gain scores between treatment and control less than or equal to .05. (Residual gain scores are calculated only for the pretest–posttest differences.)

significance in their separate relation to the control group and not in direct comparison.

The treatment effects of the three therapies are most apparent in the complaints strongly indicative of posttraumatic stress disorders, such as trauma symptoms, state anxiety, and psychoneuroticism. This points to the specificity of these forms of treatment.

Hostility	Psycho-neuroticism		Trauma symptoms		State Anxiety		State Anger	
SD	M	SD	M	SD	M	SD	M	SD
2.9	218.3	66.7	79.2	21.8	55.7	12.4	14.2	6.8
3.1	172.2[a]	65.0	56.2[ab]	24.1	45.1[ab]	13.2	12.3	6.0
3.2	171.9[ab]	73.3	55.7[ab]	26.9	41.4[ab]	14.8	12.7	6.0
5.1	241.6	54.3	85.0	16.9	58.2	10.3	12.3	3.2
5.0	194.4[a]	84.4	65.4[a]	29.4	45.0[ab]	15.7	10.9	1.9
2.6	177.2[ab]	76.4	62.0[a]	28.2	43.4[ab]	13.7	11.8	4.8
4.9	234.0	58.9	81.6	25.2	51.7	10.7	11.7	3.7
2.6	169.6[abc]	57.9	57.0[ab]	21.1	40.1[abc]	13.2	10.8	2.5
1.5	152.1[ab]	57.1	52.2[ab]	24.3	38.3[ab]	14.0	10.9	1.9
4.3	205.4	52.6	73.2	18.2	49.2	12.8	12.8	5.2
3.9	193.3	67.7	66.4	24.3	48.2	13.0	14.1	6.1

Personality

It was not our aim to bring about changes in personality with the therapeutic techniques we used. Nor did we expect any shifts in the measures that represented the various aspects of the personality. Nevertheless, some statistically significant changes in the scores can be observed. The patients consider themselves to be less distressed, and an increase in self-esteem is apparent. An even greater decrease in the score on trait anxiety indicates that in addition to the decrease of feelings of anxiety, the clients' general inclina-

Table 14.3 Personality

Therapeutic technique	Inadequacy M	Inadequacy SD	Social inadequacy M	Social inadequacy SD	Rigidity M	Rigidity SD	Discon-tentment M	Discon-tentment SD	Conceit M	Conceit SD
Trauma desensitization										
Pretest	22.7	10.1	14.6	8.8	27.7	8.3	20.1	7.6	10.6	5.1
Posttest	18.2[ab]	9.7	13.5[b]	8.6	26.3	8.5	18.9	7.9	12.4[a]	5.7
Follow-up	18.2[ab]	12.0	13.5[ab]	8.5	29.2	8.6	18.9[b]	7.9	10.8	5.5
Hypnotherapy										
Pretest	23.9	8.1	13.8	6.7	28.6	8.7	21.6	6.1	11.2	5.1
Posttest	19.3[ab]	11.1	13.6	7.2	28.7	10.4	23.0	6.8	12.4	5.8
Follow-up	16.3[ab]	10.6	11.8[b]	6.5	28.4	9.1	23.2	7.4	11.3	5.8
Psychodynamic therapy										
Pretest	25.0	8.5	13.5	8.1	25.6	8.5	20.6	7.3	11.0	5.7
Posttest	18.4[abc]	9.8	12.2[b]	7.6	24.0	8.6	22.2	13.5	11.8	6.4
Follow-up	17.5[ab]	9.8	11.2	7.3	22.6	8.8	18.3	9.3	12.8	6.3
Waiting List										
Pretest	17.2	9.4	11.3	7.1	27.3	7.0	20.7	8.1	10.9	4.0
Posttest	18.1	10.9	12.9	8.1	30.1	9.8	21.7	8.7	10.6	4.2

a *p* value of the *t* test on the difference with the pretest is less than or equal to .05.
b *p* value of the *t* test on the pretest–posttest or the pre-follow-up differences between treatment and control is less than or equal to .05.
c *p* value of the *t* test on the residual gain scores between treatment and control is less than or equal to .05. (Residual gain scores are calculated only on the pretest–posttest differences.)

tion to respond to situations with anxiety decreased. These results withstood the more rigorous testing procedures, especially in the psychodynamic therapy.

14.3 Indicators of success

Up to now, we have dealt with the treatments as if they applied to a homogeneous group of patients. The therapy, however, was beneficial to

Dominance		Self-esteem		Locus of Control		Introversion/ extroversion		Trait Anxiety		Trait Anger	
M	SD	M	SD	M	SD	M	SD	M	SD	M	SD
13.1	5.7	21.1	7.9	19.6	5.9	39.0	17.3	53.8	13.8	17.7	5.7
12.8	5.5	22.7[a]	7.8	18.8	5.6	41.7	17.0	47.2[b]	12.7	17.2	4.6
12.8	5.3	24.0[a]	8.4	19.0	7.3	41.9	18.5	47.4[ab]	15.7	17.3	5.1
12.5	6.5	22.9	7.1	22.9	4.6	44.8	15.6	57.3	10.4	21.1	5.4
13.5	7.1	23.5	8.2	22.3	5.5	42.7	16.1	45.1[abc]	16.1	20.0	6.8
13.2	7.2	25.0	9.3	21.4	5.0	44.6	21.2	45.9[ab]	13.7	18.3	5.9
12.5	6.8	22.0	6.7	19.8	6.0	47.0	7.8	57.5	10.2	20.4	4.4
16.0[bc]	7.2	25.3	7.3	18.5	6.8	50.3	18.7	45.2[abc]	10.9	18.1[a]	4.5
12.4	7.8	26.9[ab]	7.0	18.6	7.2	51.1	17.2	41.9[ab]	11.6	16.9[a]	4.0
14.1	4.7	24.7	4.6	17.2	6.5	46.5	16.0	50.4	10.8	18.6	7.8
13.2	5.8	24.1	6.4	17.4	6.8	45.0	15.2	51.4	11.3	19.9	7.3

some, whereas the situation of others remained unchanged or even became worse. The question is whether on the basis of our measures and the pretest scores, it can be predicted whether someone will or will not benefit from treatment. Are there indicators for the choice of a particular form of therapy? An explorative regression analysis was conducted whereby the residual gain scores of the variables of intrusion, avoidance and the total sum score of the Impact of Event Scale were chosen as dependent variables. These dimensions were central in the theory of the process of coping with traumatic experiences and fell within the specific objectives of the three forms of therapy. The independent variables were the personality variables, socio-demographic data, general symptoms, characteristics of the experienced event, and a few qualitative data (such as impressions of the client on the basis of the intake interview and the procedures in which the clients talked about the traumatic event). These analyses were conducted separately for each form of therapy.

From these data, we subsequently selected, once again within each form of therapy, a number of variables which proved to be instrumental in explaining the success of therapy. These variables were then jointly used in regression analyses, in order to prevent that the same variance would be explained by more than one variable.

The results indicate that different variables, used in the pretest phase, have predictive value concerning the treatment results. Moreover, it becomes clear that different predictors apply for different forms of therapy. In the case of trauma desensitization, the decrease of symptoms as measured with the Impact of Event Scale is predicted by state anger and age. This means that the treatment results are more favorable with lower pretest scores on the variable of state anger and for younger clients. In the case of trauma desensitization, the outlook for older clients and those who express anger during the pretest phase is unfavorable.

In the case of hypnotherapy, only age appears to be a predictive variable. Younger clients have a better prognosis with this form of therapy. Finally, in psychodynamic therapy, two variables repeatedly appear as predictors, namely discontentment (a mistrustful attitude towards the outside world) and locus of control. Brief psychodynamic therapy appears to work better with clients who score low on the dimension of discontentment of the Dutch Personality Questionnaire and who have low scores on the scale of internal versus external control. In other words: clients with little hostility towards others and who, in addition, feel largely in control of their own destiny, have a good chance of making progress in psychodynamic therapy of posttraumatic stress disorders.

The main factors that predict therapy results were age, locus of control, state anger, and discontentment. It must be noted that these regression analyses within the different conditions do not say anything about differential effects between the conditions. They only apply to the prognoses for clients within one condition.

That is why we conducted analyses of variance with the selected variables. In these analyses one success measure was consistently employed as a dependent variable and the type of therapy as first independent variable. We entered the selected pretest variables as second independent variable, so that the interaction effect indicated the extent to which the predictive value of the pretest variables differed between the therapies.

The most important predictor of the success of two of the three therapeutic techniques was the age of the patient. For both trauma desensitization and hypnotherapy younger patients were more likely to undergo a successful therapy, and older clients were less likely to do so. This did not apply to psychodynamic therapy (analysis of variance on the residual gain score of the overall score of the Impact of Event Scale: F therapy x age $(2, 73) = 3.6$; $p = .00$).

Another variable that indicated differential effect was the patient's income. A higher income corresponded to a better prognosis for the psychodynamic therapy and a worse prognosis for trauma desensitization. A lower income, on the other hand, predicted more positive results for trauma desensitization and a less positive result for psychodynamic therapy (F therapy x income $(2, 68) = 3.7$; $p = .02$).

Contra-indications for all three therapies were the following variables:

1. the presence of overt feelings of anger (state anger). Patients least successful during treatment were those who easily became angry, irritated and aggrieved;
2. the feeling that life is to a large extent determined by factors outside of oneself (locus of control). All therapies were more successful with those who had the feeling they were able to control their own fate. They apparently did not generalize the negative effects of the traumatic event to their life in general.

14.4 Discussion

In this chapter we reported on a controlled outcome study for posttraumatic stress disorders, disorders that are known to therapists as tenacious. For psychotherapeutic treatments specifically constructed for posttraumatic stress disorders, this study confirms that which had already been found in general evaluative studies of psychotherapy. The treatments do benefit many patients, in comparison with a control group and using stringent methodological techniques, but they do not benefit everyone to the same degree; the effects are not always substantial, and the differences between the types of therapy are small. Actually, clinically significant improvements could be observed in about 60 % of the treated patients and in 26 % of the untreated group. The similarity of the results in the three treatment conditions may be due to similarities in the behavior of the therapists, which we did not measure directly. If so, this behavior certainly is based on quite diverging theoretical considerations.

The therapeutic effects on the symptoms of intrusion and avoidance, which proved to be relatively independent dimensions, best survived the tests. This is an important finding, as these dimensions are central elements in coping with extreme stress.

Our conjecture was that short-term psychotherapy would not lead to personality changes. Some changes in the examined personality characteristics, however, did seem to take place. It is possible that a treatment consisting of 15 sessions can have an influence on some stable characteristics in the individual. A second explanation would be that the dimensions we used are more situationally specific than the literature conveys.

We would like to go into more detail concerning the differences between the three therapies. It is striking that in psychodynamic therapy the effects upon the intrusion dimension of the Impact of Event Scale clearly lag behind those on the avoidance dimension. This is just the opposite in both of the other treatment conditions; effects upon intrusion in trauma desensitization and hypnotherapy are greatest. Perhaps this result as well as the established positive aftereffects of psychodynamic therapy are specifically linked to the treatment method. Horowitz (personal communication, 1981) expects a delayed effect from psychodynamic therapy. The objective of the therapy is to get the process of coping going. In contrast to the objective of trauma desensitization, it is not so much the breaking through of the avoidance tendencies as much as it is the investigation and release of the need to avoid. Both of the other forms of therapy, most notably trauma desensitization, strive to bring about confrontations with images in order to put an end to conditioned responses. In this regard the therapy forms substantially differ from one another, and this is mirrored in the results. The clients who underwent psychodynamic therapy were usually right in the middle of a process of coping for which therapy paved the way. The message of trauma desensitization is that coping should for the most part be completed during the course of the therapy, although several coping skills are taught that could be of use at a later time.

Our findings concerning predictors of outcome provide us with interesting indications. The most important differential predictor of outcome is the age of the patient. We confirm the finding of Horowitz et al., 1984 that age is not related to outcome in psychodynamic therapy. Age is negatively correlated to outcome, however, in both trauma desensitization and hypnotherapy. Another differential predictor of outcome is the income of the patient. The nature of this variable does not provide us with a clear tool for future selection of patients. We can only assume that income is connected to self-esteem.

Besides the differential predictors we found some predictors of outcome, which are valid for the three forms of treatment alike. Contrary to the study of Abramowitz and colleagues (Abramowitz, Abramowitz, Roback & Jackson, 1974), we found that externally oriented patients have a poorer outcome, irrespective of the form of treatment. In addition, the presence of more intensive feelings of anger before therapy starts is connected to poorer outcome. This last finding can easily be interpreted within the context of the interaction between patient and therapist. It surely is more difficult in brief therapy to work with an a priori negative transference.

Finally, our findings clearly show the importance of specification of the research instruments. To continue this line, we should look for instruments that are capable of incorporating clinically relevant issues, such as the above mentioned different mechanisms within each of the therapeutic approaches. Both conclusions make it clear that the process of psychotherapy must be taken into consideration if we want to establish a more explicit link between theory, therapy, research methods, and disorders.

Epilogue

CHAPTER 15

TRAUMA IN PERSPECTIVE

In this book we have outlined a psychological approach to trauma. We have reviewed the literature and have offered an integrated perspective. There are, of course, limitations to our approach. We did not touch on many issues, such as the response of families to traumatic events, traumatization within the family, and the interaction between biological and psychological approaches. Despite these limitations, the outlined approach is valid for other fields. Our main concern, for example, was so-called type I traumas (Terr, 1991), sudden, generally brief experiences, but there is no doubt that the same general principles apply for type II traumas, repeated or prolonged exposure to extreme events.

The clinical section of this book reviews brief therapies for disorders in coping with traumatic events. The above remarks about general principles also apply to this part: there are many traumatized individuals who need long-term therapy. Clinicians, treating these patients, will undoubtedly benefit if they acquire experience in working with the treatment models that we described.

In this last chapter we formulate some notions that form a bridge between general coping theory and the treatment models.

15.1 Coping as search for meaning

Traumatic events phenomenologically have three characteristics: extreme powerlessness, disruption and extreme discomfort. All the situations discussed in this book have these characteristics in common. There is no defense against the event. The existing certainties of the individual are torn apart and a new reality forces itself upon the victim. And finally, the events cause a tremendous amount of suffering to those involved and often to those in the immediate surrounding.

The individual cannot immediately cope with such an overwhelming experience. He cannot totally ignore it either, as he is constantly reminded of it and cannot undo its occurrence. A new frame of meaning needs to be constructed in order to accept the altered situation. This applies both to grief, where the loss of a loved one has led to profound changes in a person's life, creating a need to build up a new identity as it were, and to a crime of violence, where the victim is confronted with his own vulnerability and the possibility that it could happen again.

In chapter 8, coping is considered to be a process in which the experience and its implications, on the one hand, and the existing ideas, expectations and views ('schemata'), on the other hand, are reconciled. As long as this has not occurred, the discrepancy will be continually expressed in emotions: fear of recurrence, guilt feelings over one's own responsibility, sadness over the irreversible nature of the situation, and anger because it happened to one's self.

We may conceptualize human functioning on a continuum ranging from external and observable to intra-psychic and unconscious. On one side of the continuum, we speak of behavior and of reactions to events or stimuli. The center of the continuum contains concepts such as thoughts, cognitive learning and inner imagery. The other extreme of the continuum deals with concepts even further removed from the directly observable, such as unconscious desires or intra-psychic conflicts.

Let us apply this continuum to the effects of traumatic events. We then see that the consequences of traumatic events have been most frequently described in terms of external phenomena such as symptoms and complaints. The study by Kulka et al. (1990) into the consequences of combat experiences in Vietnam, for example, focuses only on all kinds of symptoms. Reactions to serious stress situations, however, are described by others in terms of mechanisms of defense and adjustment. In this context, terms are used which refer to the intra-psychic and the unconscious, such as acute depersonalization (Noyes et al., 1977), or the lowering of awareness (Bastiaans et al., 1979) or dissociation (Van der Hart, Van der Kolk & Brown, 1989). Coping with traumatic experiences is most frequently described by such terms as intra-psychic conflicts (Horowitz, 1976) or as the release of emotional ties (Bowlby, 1980).

The past two decades have seen the emergence of the concept of coping, which allows for intra-psychic as well as behavioral ways of dealing with stress. This concept symbolizes the fundamental change in direction that has taken place in stress research over the last 15 years (Kleber, 1982). Whereas attention used to be on the causes and effects of stress, attention is now shifting to the methods people use to control tensions.

A comparable tendency can be seen in the area of coping with traumatic experiences. Instead of focusing on 'illness patterns' after traumatic events, a growing amount of research is aimed at determining how people cope with traumatic experiences. An interesting view that has surfaced in this context is that the coping process can be seen as the search for meaning. This view uses concepts located approximately at the center of the previously mentioned continuum. This approach may well offer the possibility of bridging the gap between the external and the internal extremes of the continuum.

People go through life with all kinds of images of themselves and of the world. The assumption of invulnerability is important for functioning adequately (Janoff-Bulman, 1989). We do not provide for the possibility of accidents, crimes or the loss of a loved one. The 'nothing will happen to me' attitude enables us to undertake many activities without fear, which, when subjected to closer examination, are more perilous than we are inclined to believe.

Another deeply anchored view is that the world functions in an understandable and orderly fashion (Janoff-Bulman & Frieze, 1983). We view our world as being controllable to a certain degree. Personal disasters will not overcome us because we are too cautious to let them occur.

A third general condition under which people function is a positive self-image. People generally see themselves as decent and valuable persons (McCann & Pearlman, 1990).

When people fall victim to traumatic events such as crimes of violence, traffic accidents or the sudden loss of a loved one, these fundamental assumptions surface. Why did this have to happen to me? This commonly voiced question illustrates the degree of doubt regarding all three of the above mentioned 'certainties'. How can one be sure that something similar will not happen again? The ideas on which our sense of security and certainty are based have abruptly been overturned. Invulnerability is clearly an illusion; the world is not so predictable and orderly, and the victim is confronted with all kinds of negative self-images.

The task that victims need to accomplish is the restoration of a view of the world and of themselves that will enable them to feel reasonably safe and secure again. This process of restoration can take place in different ways.

One possibility is the isolation of the event by denying its emotional meaning. This mechanism appears to be used frequently in the case of traffic accidents. The motto here is ' back behind the wheel as soon as possible, and thank your lucky stars'. This enables the maintenance of a sense (or illusion) of invulnerability and control of the situation.

Another way of coping is cognitive framing. This is probably one of the most important coping methods. After crimes of violence, for instance, people develop a pattern of signs; signals intended to warn them for a possible repeat of the event (chapter 5). A well-known example of such a signal is the scene of the crime. This location evokes anxiety and is often avoided. Not only is this an instance of conditioning, but such a signal can also be viewed as a curtailment of anxiety. The scene of the crime is a dangerous place; other locations are safe.

There may also be different internal restrictions attached to the meaning of the event. An example in this context is the increased sense of self-blame concerning the event. The guilt feelings manifested by parents who have lost their infant through Sudden Infant Death Syndrome could partly be intended to decrease the fear of it happening to their next child. These reactions are common to all people, albeit in less dramatic circumstances, such as reexperiencing a certain situation and only then realizing what one should have said, or reconstructing a traffic accident and seeing what should have been done to avoid it. These are forms of 'secondary control' (Rothbaum et al., 1982), that is to say people attempt to harmonize their own experience with the environment.

15.2 Normal coping

The approach we have outlined in this book does not make a qualitative distinction between normal coping and disorders. A psychological view of the consequences of traumatic events does not consider psychiatric diagnoses as a logical consequence of coping. The concepts of psychiatry do not fit the psychological model of coping. They draw an arbitrary line in a continuum that ranges from a relatively short process of coping, in which people feel quite able to deal with the consequences of the event, to a chronic, frozen state either of being overwhelmed by feelings, thoughts and reminiscences of the event or of numbness and masked depression as a sign of the inability to cope with the event.

We stress here the importance of looking at the process of coping with traumatic events as a psychological process. This perspective opens up possibilities, which remain closed when one only looks at disorders. Both in general health care and in mental health care there are many people who apply for help after serious events, but who do not necessarily suffer from psychiatric disorders. It is clear that these victims have been neglected in mental health care. The adherence to dichotomous diagnostics is one of the causes of this neglect (Boulanger, 1990; Davidson & Smith, 1990).

Based on the perspective discussed above, the possibility to prevent disorders by early intervention in the coping process is a logical consequence. The approach we described in chapter 10 was developed as a response to the persisting complaint of victims that the environment was not willing or able to listen to the experiences they wanted to share. Even during the implementation of prevention programs (that are executed by professionals) a common finding is that health professionals or personnel consultants can more easily detect specific and defined pathology than they are willing to listen to the details of traumatic events. To ensure adequate care for victims in organizations, much organizational work has to be done. Suggestions for implementation in organizations are also given in chapter 10.

15.3 Psychotherapy and attribution of meaning

Psychotherapy can be viewed as the presentation of new structures of meaning. Client and therapist develop activities that are explicitly or implicitly aimed at yielding new feelings and insights for the client. The meaning, or rather often the lack of meaning the client ascribes to his problems is changed by the interaction with the therapist. The way in which this is done differs strongly from one form of therapy to another. We will now consider the three forms of therapy used in our study especially in terms of the way in which they promote the assignation of meaning by the client.

It should be noted that the frameworks and meanings described are very general. Every individual experiences these or other themes and conflicts in his own particular way. In this context we wish to emphasize that only a few examples have been given here, and that only some characteristics of the various forms of therapy will be considered.

Which frameworks relevant to adequate coping are offered by trauma desensitization? How are the event and the symptoms stemming from it viewed in this method? The message that trauma desensitization (chapter

11) gives to the client is aimed at reality-testing. He is (implicitly or explicitly) told that his problems are related to the avoidance of memories of the event. He can regain control over his complaints by practicing relaxation exercises and by consciously confronting previously avoided stimuli. Recoding, in the sense that negative events are definitely history, adds to regaining security and future perspective.

Through trauma desensitization, the meaning of the traumatic experience is focused on the manner in which the client deals with the event and with his symptoms. The only way in which a link can be made with the client's general functioning is through the question of why confrontation was avoided for so long.

Trauma desensitization enables the client to actively regain the feeling of having control over his life. On the one hand, he achieves this through relaxation, which promotes the realization that he is less dependent upon the environment for having positive experiences. On the other hand, he works through the experience in a renewed, gradual confrontation, which he himself constantly starts and ends.

Hypnotherapy (chapter 13) cannot be viewed as one single approach. Any form of therapy, whatever its theoretical lineage, can be referred to as hypnotherapy if hypnosis forms a part of it. In our study, learning theory formed the background of the treatment. An important difference with trauma desensitization, however, is that the process of treatment is directed through the client's imagery. In hypnotherapy, the therapist aims his efforts at the client's metaphors, whereas in trauma desensitization it is the therapist who directs the client's imagination. It is difficult to describe just exactly what clients go through during trance experiences. There is an impression that it is not only the traumatic experience that is relived in the metaphors, but also the personality problems that lie at the basis of the disorder.

At first glance, our form of hypnotherapy appears to offer a similar framework of meaning to that of trauma desensitization. However, if we consider involving broader topics in the treatment, then entirely different meanings could come into the picture. The lack of theory formation concerning hypnotherapy, however, has its effect here. As long as it remains unclear what status we should assign to the use of hypnosis (is it merely a relaxation technique or more than that?), it also remains unclear what we are offering the client. The chance that certain factors and preferences linked to the individual therapist will play an important role should not be underestimated.

Psychodynamic therapy (chapter 12), ideally, does not limit itself to the recent traumatic experience alone. The present conflict is placed within the framework of personal development and the way in which the client interacts with others. In practice, however, the way in which the focus is handled is dependent on an estimation of the client's capabilities. When the client has a good capacity for integration and a past that is relatively unburdened, the therapist will attempt to link both the coping process and the transference reactions to subjects that were important to the individual's development. In the case of clients with a less stable background, the interpretations will be much more closely linked to the actual traumatic experience.

What can be said about the similarities and differences between the forms of therapy in terms of the indicated perspective? One point of similarity is that all three methods offer a framework in which the reaction to the traumatic event can be placed and understood by the client. The ways in which changes in the assignation of meaning are achieved, however, differ greatly. We can ask ourselves to what extent these methods are equal in this respect. Frankl (1959), for example, indicates that any meaning which gives strength to the client is acceptable. Taylor (1983) confirms that 'a' meaning is better than no meaning at all. She found that people who assign a certain meaning to their suffering feel better than people who cannot attach any meaning to it. In conclusion, we can argue that the methods of treatment studied facilitate the coping process by offering a framework.

One difference between the various therapies concerns the range of the meaning offered. It concerns the question of which aspects of life can be included in the framework. The framework offered by trauma desensitization is strictly defined. The event and the reaction to it are basically separate from the rest of the way an individual functions. The experience is, in fact, turned into an isolated phenomenon, which can be dealt with and left behind.

Psychodynamic psychotherapy, and, to a lesser extent, hypnotherapy, involve a much broader focus of treatment. The event and the reaction to it are placed in the context of the development and interactional style of the client. In contrast to the assignation of a meaning, the 'underlying' meanings are disentangled and placed into a personal context.

The different therapies are evidently suited to certain individuals under certain conditions. The approach outlined above can, when elaborated upon and studied, serve as an indicator to match patient and treatment. This could be achieved, particularly through process research, in which the process of coping with a traumatic event is studied in terms of a perspective of the individual attribution of meaning. Such research will clarify how specific ways of meaning attribution contribute to the coping process.

15.4 The future of the trauma area

Since the beginning of the seventies the interest in the consequences of traumatic events has been increasing both in science and in society in general. Social factors such as the war in Vietnam and the growing care for victims of violence have contributed to this development. The attention of the scientific world for the processes of coping is part of a broader range of research on stress with an accent on the transactional approach (Lazarus, 1981). The dynamic character of coping and the importance of cognitive aspects have become central issues. Coping with serious events can be considered as a process in which people attempt to regain their fundamental sense of trust.

In the future we expect the following developments:

1. Psychological research will focus on normal coping, in contrast to psychiatric research, in which the search for the biological correlates of disorders is prevailing at present. Until now there has been hardly any longitudinal research on representative samples of people who went through specific distressing events. This research will have to contribute to the theory on normal and pathological ways of coping.
2. Another research issue will be the factors which facilitate or impede coping. Only some factors are known until now. The determination of risk factors and their relative influence will become useful tools for planning and implementing mental health care.
3. Finally, victim support work will go in the direction of preventive counselling. More and more the accent will be on early recognition and prevention of disorders, while therapies will be attuned more specifically to the disorder which interferes with normal life.

REFERENCES

Abrahams, M.J., Price, J., Whitlock, F.A. & Williams, G. (1976). The Brisbane floods, January 1974: Their impact on health. *The Medical Journal of Australia, 2*, 936-939.

Abramowitz, C.V., Abramowitz, S.I., Roback, H.B. & Jackson, C. (1974). Differential effectiveness of directive and nondirective group therapies as a function of client internal-external control. *Journal of Consulting and Clinical Psychology, 42*, 849-853.

Abramson, L.Y., Seligman, M.E.P., & Teasdale, J.D.(1978). Learned helplessness in human: Critique and reformulation. *Journal of Abnormal Psychology, 87*, 49-74.

Adams, P.R. & Adams, G.R. (1984). Mount Saint Helen's Ashfall: Evidence for a disaster stress reaction. *American Psychologist*, 39, 252-260.

Alexander, F. (1950). *Psychosomatic Medicine.* London: Allen and Unwin.

Alexander, F. & French, T. M. (1946). *Psychoanalytic therapy.* New York: Ronald Press Co.

American Psychiatric Association (1968). *Diagnostic and statistical manual of mental disorders, second edition (DSM-II).* Washington, D.C.: A.P.A.

American Psychiatric Association (1980). *Diagnostic and statistical manual of mental disorders, third edition (DSM-III).* Washington, D.C.: A.P.A.

American Psychiatric Association (1987). *Diagnostic and statistical manual of mental disorders, third edition revised (DSM-III-R).* Washington, D.C.: A.P.A.

Anderson, C.R. (1976). Coping behaviors as intervening mechanisms in the inverted-U stress-performance relationship. *Journal of Applied Psychology, 61*, 30-34.

Anderson, C.R. (1977). Locus of control, coping behaviors, and performance in a stress setting: A longitudinal study. *Journal of Applied Psychology, 62*, 446-451.

Andreasen, N.J.C., Norris, A.S. & Hartford, C.R. (1971). Incidence of long-term psychiatric complications in severely burned adults. *Annals of Surgery, 174*, 785-793.

Andriessen, J.H.T.H. (1972). Interne of externe beheersing. *Nederlands Tijdschrift voor de Psychologie, 27*, 173-199.

Anisman, H. & Zacharko, R.M. (1982). Depression: The predisposing influence of stress. *The Behavioral and Brain Sciences, 5*, 89-137.

Antonovsky, A., Maoz, B., Dowty, N. & Wijsenbeek H. (1971). Twenty-five years later: A limited study of the sequelae of the concentration camp experience. *Social Psychiatry, 6*, 186-193.

Archibald, H.C., Long, D.M., Miller, C. & Tuddenham, R.D. (1962). Gross stress reactions in combat: A 15 years follow-up. *American Journal of Psychiatry, 119,* 317-322.

Archibald, H.C. & Tuddenham, R.D. (1965). Persistent stress reactions after combat. *Archives of General Psychiatry, 12,* 475-481.

Ariès, P. (1974). *Western attitudes toward death: From the middle ages to the present.* Baltimore/London: The Johns Hopkins University Press.

Arindell, W. A. & Ettema, H. (1981). Dimensionele structuur, betrouwbaarheid en validiteit van de Nederlandse bewerking van de Symptom Checklist (SCL-90). *Nederlands Tijdschrift voor de Psychologie, 36,* 77-108.

Averill, J.R. (1968). Grief: Its nature and significance. *Psychological Bulletin, 70,* 721-748.

Averill, J.R. (1979a). The functions of grief. In C. Izard (Ed.), *Emotions in personality and psychopathology* (pp. 339-368). New York: Plenum.

Averill, J.R. (1979b). A selective review of cognitive and behavioral factors involved in the regulation of stress. In R.A. Depue (Ed.), *The psychobiology of the depressive disorders: Implications for the effects of stress.* New York: Academic Press.

Baker, G.W. & Chapman, D.W. (1962). *Man and society in disaster.* New York: Basic Books.

Bally, G. (1969). *De psychoanalyse van Sigmund Freud.* Utrecht/Antwerpen: Het Spectrum.

Bard, M. & Sangrey, D. (1979). *The crime's victim book.* New York: Basic Books.

Barocas, H. & Barocas, C. (1980). Separation - individuation conflicts in children of holocaust survivors. *Journal of Contemporary Psychotherapy, 11,* 6-14.

Barrett, T.W. & Mizes, J.S. (1988). Combat level and social support in the development of posttraumatic stress disorder in Vietnam veterans. *Behavior Modification, 12,* 110-115.

Barton, A.H. (1969). *Communities in disaster: A sociological analysis of collective stress situations.* New York: Double Day.

Bastiaans, J. (1957). *Psychosomatische gevolgen van onderdrukking en verzet.* Amsterdam: Noord-Hollandse Uitgevers Maatschappij

Bastiaans, J. (1970). Over de specificiteit en de behandeling van het KZ-syndroom. *Nederlands Militair Geneeskundig Tijdschrift, 23,* 364.

Bastiaans, J. (1974). Het KZ-syndroom en de menselijke vrijheid. *Nederlands Tijdschrift voor Geneeskunde, 118,* 1173-1178.

Bastiaans, J., Jaspers, J.P.C., Ploeg, H.M. van der, Berg- Schaap, Th.E. van den & Berg, J.F. van den (1979). *Psychologisch onderzoek naar de gevolgen van gijzelingen in Nederland. (1974-1977).* 's-Gravenhage: Staatsuitgeverij.

Baum, A., Fleming, R. & Singer, J.E. (1983). Coping with victimization by technological disaster. *Journal of Social Issues, 39,* 117-138.

Baum, A. & Singer, J.E. (1982). Psychosocial aspects of health, stress and illness. In A.H. Hastdorf & A.M. Isen (Eds.), *Cognitive Social Psychology* (pp. 307-355). New York: Elsevier North-Holland.

Beal, S.M. (1979). Sudden infant death syndrome. *Australian Family Physician, 8,* 1279-1283.

Beecher, H.K. (1956). Relationship of significance of wound to pain experienced. *Journal of the American Medical Association, 161,* 1609-1613.

Begemann, F.A. (1991). *Het onvertelbare: een verkennend onderzoek naar psychotherapie met oorlogsgetroffenen en hun kinderen.* Amsterdam/Lisse: Swets & Zeitlinger.

Bennett, G. (1970). Bristol floods 1968: A controlled survey of effects on health of local community disaster. *British Medical Journal, 3,* 454.

Berkman, L.F. & Syme, S.L. (1979). Social networks, lost resistence, and mortality: A nine-year follow-up study of Alameda county residents. *American Journal of Epidemiology, 109,* 186

Berren, M.R., Beigel, A. & Ghertner, S. (1980). A typology for the classification of disasters. *Community Mental Health Journal, 16,* 103-111.

Bettelheim, B. (1943). Individual and mass behavior in extreme situations. *Journal of Abnormal and Social Psychology, 38,* 417-452.

Blanchard, E. B., & Abel, G. (1976). An experimental case study of the biofeedback treatment of a rape-induced psychophysiological cardiovascular disorder. *Behavior Therapy, 7,* 113-119.

Bloom, B.L., White, S.W. and Asher, S.J. (1978). Marital disruption as a stressful life event. *Psychological Bulletin, 85,* 867-894.

Boeke, P.E. (1984). *De prinses op de erwt?* Groningen, The Netherlands: University of Groningen, valedictory lecture.

Boman, B. (1979). Behavioural observations on the Granville train disaster and the significance of stress for psychiatry. *Social Science and Medicine, 13A,* 463-471.

Boman, B. (1986). Early experiential environment, maternal bonding and the susceptibility to posttraumatic stress disorder. *Military Medicine, 151,* 528-531.

Boudewyns, P.A., Hyer, L., Woods, M.G., Harrison, W.R. & McCranie, E. (1990). PTSD among Vietnam veterans: An early look at treatment outcome using direct therapeutic exposure. *Journal of Traumatic Stress, 3,* 359-368.

Boulanger, G. (1990). A state of anarchy and a call to arms: The research and treatment of post-traumatic stress disorder. *Journal of Contemporary Psychotherapy, 20,* 5-15

Bourne, P.G. (1970). Military psychiatry and the Vietnam experience. *American Journal of Psychiatry, 127,* 123-130.

Bowen, G. R. & Lambert, J. A. (1985). Systematic desensitization therapy with posttraumatic stress disorder cases. In C. R. Figley (Ed.), *Trauma and its wake.* New York: Brunner/Mazel.

Bowlby, J. (1961). Processes of mourning. *International Journal of Psycho-Analysis, 42,* 317-340.

Bowlby, J. (1980). *Attachment and loss (Volume III) - Loss: Sadness and depression.* London: The Hogarth Press.

Braverman, M. (1980). Onset of psychotraumatic reactions. *Journal of Forensic Sciences, 25,* 821-825.

Brende, J.O. (1987). Dissociative disorders in Vietnam combat veterans. *Journal of Contemporary Psychotherapy, 17,* 77-86.

Brende, J.O. & Benedict, B.D. (1980). The Vietnam combat delayed stress response syndrome: Hypnotherapy of "dissociative symptoms". *The American Journal of Clinical Hypnosis, 23,* 34-40.

Brenner, C. (1981). Defense and defense mechanisms. *Psychoanalytic Quarterly, 50,* 557-569.

Brett, E.A. & Ostroff, R. (1985). Imagery and posttraumatic stress disorder. *American Journal of Psychiatry, 142,* 417-424.

Breuer, J. & Freud, S. (1895/1952). *Studien über Hysterie. Gesammelte Werke I.* London: Imago Publishing Company.
Brill, N.Q. (1967). Gross stress reaction. II: Traumatic war neurosis. In A.M. Freedman & H.I. Kaplan (Eds.), *Comprehensive textbook of psychiatry* (pp. 1031-1035). New York: Williams and Williams.
Brill, N.Q. & Beebe, B.W. (1955). *A follow-up study of war neuroses.* Washington, D.C.: Veterans Administration Medical Monograph.
Brom, D. & Kleber, R.J. (1985). De Schok Verwerkings Lijst. *Nederlands Tijdschrift voor de Psychologie, 40,* 164-168.
Brom, D., Kleber, R.J. & Defares, P.B.(1986). *Traumatische ervaringen en psychotherapie.* [Traumatic experiences and psychotherapy]. Lisse: Swets & Zeitlinger.
Brom, D. & Kleber R.J. (1988). Psychotherapy and pathological grief: I. The search for meaning. In E. Chigier (Ed.), *Grief and Bereavement in Contemporary Society* (pp. 52-58). London: Freund Publishing House, Volume 2.
Brom, D. & Kleber, R.J. (1989). Prevention of posttraumatic stress disorders. *Journal of Traumatic Stress Studies, 2,* 335-351.
Brom, D., Kleber, R.J. & Defares, P.B. (1989). Brief psychotherapy for posttraumatic stress disorders. *Journal of Consulting and Clinical Psychology, 57,* (5), 607-612.
Brown, G.W. (1979). A three-factor causal model of depression. In J.E. Barrett, R.M. Rose & G.L. Klerman (Eds.), *Stress and mental disorder* (pp. 111-121). New York: Raven Press.
Brown, G.W. (1981). Contextual measures of life events. In B.S. Dohrenwend & B.P. Dohrenwend (Eds.), *Stressful life events and their contexts* (pp. 187-201). New York: Prodist.
Brown, G.W., Davidson, S., Harris, T., Maclean, V., Pollack, S. & Prudo, S. (1977). Psychiatric disorder in London and North Uist. *Social Science and Medicine, 11,* 367.
Buelens, J. (1971). *Sigmund Freud: kind van zijn tijd.* Meppel: Boom.
Bulman, R.J. & Wortman, C.B. (1977). Attributions of blame and coping in the "real world": Severe accident victims react to their lot. *Journal of Personality and Social Psychology, 35,* 351-363.
Burgess, A.W. & Holmstrom, L.L. (1974). Rape trauma syndrome. *American Journal of Psychiatry, 31,* 981-986.
Burgess, A.W. & Holmstrom, L.L. (1978). Recovery from rape and prior life stress. *Research in Nursing and Health, 1,* 165-174.
Burnell, G.M. & Burnell, A.L. (1989). *Clinical management of bereavement: A handbook for health care professionals.* New York: Human Sciences Press.
Caplan, G. (1964). *Principles of preventive psychiatry.* New York: Basic Books.
Card, J.J. (1987). Epidemiology of PTSD in a national cohort of Vietnam veterans. *Journal of Clinical Psychology, 43,* 6-17.
Carey, R.G. (1977). The widowed: A year later. *Journal of Counseling Psychology, 24,* 125-131.
Cassee, A. P., Petrie, J. F, & De Wolf, J. (1984). Enkele ervaringen met kortdurende psychodynamische therapie. *Tijdschrift voor Psychotherapie, 10,* 217-231.
Catanese, C.A. (1978). *Case studies of the post-traumatic effects of the PAN-AM - KLM aviation disaster in the Canary Islands.* Ann Arbor: University Microfilms International, Doctoral dissertation.

Catherall, D.R. (1989). Differentiating intervention strategies for primary and secondary trauma in posttraumatic stress disorder: The example of Vietnam veterans. *Journal of Traumatic Stress, 2*, 289-304.

Chapman, D.W. (1962). A brief introduction to contemporary disaster research. In G.W. Baker & D.W. Chapman (Eds.), *Man and society in disaster* (pp. 3-22). New York: Basic Books.

Charmaz, K. (1980). *The social reality of death*. Reading, Mass.: Addison-Wesley.

Chertok, L. (1984) Hypnosis and suggestion in a century of psychotherapy: An epistemological assessment. *Journal of the American Academy of Psychoanalysis, 12*, 211-232.

Chodoff, P. (1966). Extreme coercive and oppressive forces: Brainwashing and concentration camps. In S. Arieti (Ed.), *American Handbook of Psychiatry, volume III*. New York: Basic Books.

Clayton, P.J. (1975). The effect of living alone on bereavement symptoms. *American Journal of Psychiatry, 132*, 133-137.

Cleary, P.D. & Houts, P.S. (1984). The psychological impact of the Three Mile Island incident. *Journal of Human Stress, 10*, 28-34.

Cohen, E.A. (1953). *Human behavior in the concentration camp*. New York: W.W. Norton and Company. (original Dutch edition: 1952).

Cohen, L. & Roth, S. (1984). Coping with abortion. *Journal of Human Stress, 10*, 140-145.

Cohen, R.E. (1976). Postdisaster mobilization of a crisis intervention team: The Managua experience. In H.J. Parad, H.L.P. Resnik & L.G. Parad (Eds.), *Emergency and disaster management* (pp. 375-385).

Cohen, R.E. & Ahearn, F.L. (1980). *Handbook for mental health care of disaster victims*. Baltimore: Johns Hopkins University Press.

Culpan, R. & Taylor, C. (1973). Psychiatric disorders following road traffic and industrial injuries. *Australian and New Zealand Journal of Psychiatry, 7*, 1-7.

Danieli, Y. (1982). Families of survivors of the Nazi Holocaust: Some short and long term effects. In C.D. Spielberger & I.G. Sarason (Eds.), *Stress and Anxiety, Volume 8* (pp. 405-421). New York: Hemisphere.

Dasberg, H. (1987). Psychological distress of Holocaust survivors and offspring in Israel, forty years later: A review. *Israel Journal of Psychiatry and Related Sciences, 24*, 243-256.

Davidson, J. & Smith, R. (1990). Traumatic expriences in psychiatric outpatients. *Journal of Traumatic Stress, 3*, 459-475.

Davidson, L.M. & Baum, A. (1986). Chronic stress and posttraumatic disorders. *Journal of Clinical and Consulting Psychology, 54*, 303-308.

Defares, P.B. (1982). *Description of the ventilatory feedback training device*. Wageningen, The Netherlands: Institute of Stress Research.

Defares, P.B. (1990). Determinants of anxiety changes. In C.D. Spielberger & I.G. Sarason (Eds.), *Stress and Anxiety, volume 13* (pp. 117-130). Washington, D.C.: Hemisphere.

Defares, P.B. & Grossmann, P. (1988). Hyperventilation, stress and health risk behavior. In S. Maes, P.B. Defares & I.G. Sarason (Eds.), *Topics in health psychology* (pp. 141-156). Chichester: Wiley.

Defares, P.B., Brandjes, M., Nass, C.H.Th. & Van der Ploeg, J.D. (1985). Coping styles, social support and sex-differences. In I.G. Sarason & B.R. Sarason (Eds.), *Social support: Theory, research and applications* (pp. 173-187). Dordrecht: Martinus Nijhoff Publishers.

DeFazio, V.J. (1975). The Vietnam era veteran: Psychological problems. *Journal of Contemporary Psychotherapy, 7*, 9-15.

DeFazio, V.J. (1978). Dynamic perspectives on the nature and effects of combat stress. In C.R. Figley (Ed.), *Stress disorders among Vietnam veterans: Theory, research and treatment*. New York: Brunner/Mazel.

De Frain, J.D. & Ernst, L. (1978). The psychological effects of sudden infant death syndrome on surviving family members. *The Journal of Family Practice, 6*, 985-989.

De Graaf, Th. (1975). Pathological patterns of identification in families of survivors of the Holocaust. *Israel Annals of Psychiatry and Related Disciplines, 13*, (4), 335-363.

Dennett, D.C. (1980). *Brainstorms*. Hassocks: Harvester Press.

De Wind, E. (1949). Confrontatie met de dood. *Folia Psychiatrica, Neurologica et Neurochirurgica Neerlandica, 52*, 459-466.

Dimsdale, J.E. (1974). The coping behavior of Nazi Concentration camp survivors. *American Journal of Psychiatry, 131*, 792-797

Doreleijers, T.A.H. & Donovan, D.M. (1990). Transgenerational traumatization in children of parents interned in Japanese civil internment camps in the Dutch Indies during World War II. *The Journal of Psychohistory, 17*, 435-447.

Dohrenwend, B.S. & Dohrenwend, B.P. (1981). Life stress and illness: Formulation of the issues. In B.S. Dohrenwend & B.P. Dohrenwend (Eds.), *Stressful life events and their contexts* (pp. 1-27). New York: Prodist.

Dor-Shav, N.K. (1978). On the long-range effects of concentration camp internment on Nazi victims: 25 years later. *Journal of Consulting and Clinical Psychology, 46*, 1-11.

Duyker, H.C.J. (1972). De meervoudige gedetermineerdheid van gedrag. *Mededelingen der Koninklijke Nederlandse Academie van Wetenschappen, afdeling Letterkunde, 35*, 3.

Dynes, R.R. & Quarantelli, E.L. (1976). The family and community context of individual reactions to disaster. In H.J. Parad, H.L.P. Resnik & L.G. Parad (Eds.), *Emergency and Disaster Management* (pp. 231-245).

Eissler, K. (1963). Die Ermordung von wievielen seiner Kinder muss ein Mensch symptomfrei ertragen konnen um eine normale Konstitution zu haben? *Psyche, 1*, 241.

Eitinger, L. (1964). *Concentration camp survivors in Norway and Israel*. Oslo/London: Oslo Universitetsforlaget/ Allen and Unwin.

Eitinger, L. (1980). The concentration camp syndrome and its late sequelae. In J.E. Dimsdale (Ed.), *Survivors, victims and perpetrators: Essays on the Nazi Holocaust* (pp. 127-162). Washington, D.C.: Hemisphere, 127-162.

Eitinger, L. & Strøm, A. (1973). *Mortality and morbidity after excessive stress*. Oslo: Universitetsforlaget/ New York: Humanities Press.

Eland, J., Van der Velden, P.G., Kleber, R.J. & Steinmetz, C.H.D. (1990). *Tweede generatie Joodse Nederlanders: een onderzoek naar gezinsachtergronden en psychisch functioneren*. Deventer: Van Loghum Slaterus.

Ellemers, J.E. (1956). *De Februari-ramp*. Assen: Van Gorcum.

Endler, N.S. & Magnusson, D. (1976). Toward an international psychology of personality. *Psychological Bulletin, 83,* 956-974.

Engel, G.L. (1961). Is grief a disease? *Psychosomatic Medicine, 23,* 18-22.

Erichsen, J.E. (1866). *On railway spine and other injuries of the nervous system*. London: Walton and Maberly.

Erickson, M.H. (1952). Deep hypnosis and its induction. In L.M. LeCron (Ed.), *Experimental hypnosis*. New York: Macmillan.

Erikson, K.T. (1976). *Everything in its path: Destruction of community in the Buffalo Creek Flood*. New York: Simon and Schuster.

Eth, S. & Pynoos, R.S. (1985) (Eds.). *Post-traumatic stress disorder in children*. Washington: American Psychiatric Press.

Evans, H.I. (1978). Psychotherapy for the rape victim: Some treatment models. *Hospital and Community Psychiatry, 29,* (5), 309-312.

Evans, I.M. (1973). The logical requirements for explanations of systematic desensitization. *Behavior Therapy, 4,* 506-514.

Ewalt, J.R. (1981). What about the Vietnam veteran? *Military Medicine, 146,* 3.

Eysenck, H.J. (1980). Psychological theories of anxiety. In G.D. Burrows & B. Davies (Eds.), *Handbook of studies on anxiety*. Amsterdam: Elsevier/North Holland Biomedical Press.

Fairbank, J.A. & Brown, T. (1987). Current behavioral approaches to the treatment of posttraumatic stress disorder. *Behavior Therapy, 3,* 57-64.

Fairbank, J.A. & Keane, T.M. (1982). Flooding for combat-related stress disorders: Assessment of anxiety reduction across traumatic memories. *Behavior Therapy, 13,* 499-510.

Fairbank, J.A., Keane, T.M. & Malloy, P.F. (1983). Some preliminary data on the psychological characteristics of Vietnam veterans with posttraumatic stress disorders. *Journal of Consulting and Clinical Psychology, 51,* 912-919.

Fairbank, J.A., Langley, M.K., Jarvie, G.J. & Keane, T.M. (1981). A selected bibliography on posttraumatic stress disorders in Vietnam veterans. *Professional Psychology, 12,* 578-586.

Fenichel, O. (1945). *The psychoanalytic theory of neurosis*. New York: Norton.

Ferenczi, S., & Rank, O. (1926). *The development of psychoanalysis*. New York: Nervous and Mental Diseases Publishing Company.

Figley, C.R. (1978). Symptoms of delayed combat stress among a college sample of Vietnam veterans. *Military Medicine, 143,* 107-110.

Finkel, N.J. (1976). *Mental illness and health: Its legacy, tensions, and changes*. New York: MacMillan Publishing Company.

Flannery, R.B. (1990). Social support and psychological trauma: A methodological review. *Journal of Traumatic Stress, 3,* 593-613.

Foeckler, M.M., Garrard, F.H., Williams, C.C., Thomas, A.M. & Jones, T.J. (1978). Vehicle drivers and fatal accidents. *Suicide and Life-threatening Behavior, 8,* 174-182.

Folkman, S. & Lazarus, R.S. (1985). If it changes it must be a process: Study of emotion and coping during three stages of a college examination. *Journal of Personality and Social Psychology, 48,* 150-170.

Foy, D.W., Carroll, E.M. & Donahoe, C.P. (1987). Etiological factors in the development of posttraumatic stress disorder in clinical samples of Vietnam combat veterans. *Journal of Clinical Psychology, 43,* 17-27.

Frank, E., Turner, S.M. & Duffy, B. (1979). Depressive symptoms in rape victims. *Journal of Affective Disorders, 1*, 269-277.

Frankl, V.E. (1959). *Man's search for meaning: An introduction to logotherapy*. Boston: Beacon Press.

Frederick, C.J. (1980). Effects of natural versus human-induced violence upon victims. *Evaluation and Change, special issue*, 71-75.

French, J.R.P. & Kahn, R.L. (1962). A programmatic approach to studying the industrial environment and mental health. *Journal of Social issues, 18*, 1-47.

Freud, S. (1917/1985). *Trauer und Melancholie*. Meppel: Boom/Freud, Psychoanalytische Theorie 1 (Dutch edition).

Freud, S. (1917/1940). *Vorlesungen zur Einführung in die Psychoanalyse. Gesammelte Werke XI*. London: Imago Publishing Company.

Freud, S. (1919/1947). *Einleitung zu Zur Psychoanalyse der Kriegsneurosen. Gesammelte Werke XII*. London: Imago Publishing Company.

Freud, S. (1920/1955). *Jenseits des Lustprinzips. Gesammelte Werke XIII*. London: Imago Publishing Company.

Freud, S. (1935/1955). *Abriss der Psychoanalyse. Gesammelte Werke XVII*. London: Imago Publishing Company.

Freyberg, J.T.(1980). Difficulties in separation-individuation as experienced by offspring of Nazi Holocaust survivors. *American Journal of Orthopsychiatry, 50*, 87-95.

Friedman, P. (1949). Some aspects of concentration camp psychology. *American Journal of Psychiatry, 105*, 601-605.

Friedman, P. & Linn, L. (1957). Some psychiatric notes on the 'Andrea Doria' disaster. *American Journal of Psychiatry, 114*, 426.

Friedsam, H.J. (1962). Other persons in disaster. In G.W. Baker & D.W. Chapman (Eds.), *Man and society in disaster* (pp. 151-182). New York: Basic Books.

Frijda, N.H. (1982). Over angst en depressie. *Tijdschrift voor Psychotherapie, 8*, 245-256.

Frijda, N.H. (1982). *The Emotions*. Cambridge: Cambridge University Press.

Fritz, C.E. & Marks, E.A. (1954). The NORC studies of human behavior in disaster. *Journal of Social Issues, 10*, 26-41.

Fromm, E. (1984). Hypnoanalysis with particular emphasis on the borderline patient. *Psychoanalytic Psychology, 1*, 61-76.

Fromm, E. & Hurt, S.W. (1980). Ego-psychological parameters of hypnosis and other altered states of consciousness. In G.D. Burrows & L. Dennerstein (Eds.), *Handbook of hypnosis and psychosomatic medicine*. Amsterdam: Elsevier/North-Holland.

Fromm-Reichmann, F. (1942/1959). *Psychoanalysis and psychotherapy: Selected papers*. Chicago: The University of Chicago Press.

Frye, J.S. & Stockton, R.A. (1982). Discriminant analysis of posttraumatic stress disorder among a group of Vietnam veterans. *American Journal of Psychiatry, 139*, 52-56.

Furst, S. (Ed.) (1967). *Psychic trauma*. New York: Basic Books.

Garfield, S.L. (1978). Research on client variables in psychotherapy. In S.L. Garfield & Bergin, A.E., *Handbook of psychotherapy and behavior change: An empirical analysis*. New York: Wiley.

Gersons, B.P.R. (1989). Patterns of PTSD among police officers following shooting incidents: A two-dimensional model and treatment implications. *Journal of Traumatic Stress, 2*, 247-258.

Giel, R. (1991). The psychosocial aftermath of two major disasters in the Soviet Union. *Journal of Traumatic Stress, 4*, 381-392.

Gilligan, S.G. (1981). Ericksonian approaches to clinical hypnosis. In J.K. Zeig (Ed.), *Ericksonian psychotherapy*. New York: Brunner/Mazel.

Glass, A.J. (1954). Psychotherapy in the combat zone. *Journal of Psychiatry, 110*, 725-731.

Glick, I.O., Weiss, R.S. & Parkes, C.M. (1974). *The first year of bereavement*. New York: Wiley Interscience.

Goldberg, A. (1973). Psychotherapy of narcissistic injuries. *Archives of General Psychiatry, 28*, 722-726.

Goldfried, M.R. (1971). Systematic desensitization as training in self-control. *Journal of Consulting and Clinical Psychology, 37*, 228-234.

Goldney, R.D. (1981). Parental loss and reported childhood stress in young women who attempt suicide. *Acta Psychiatrica Scandinavica, 64*, 34-59.

Gorer, G. (1965). *Death, grief and mourning in contemporary Britain*. London: Tavistock Publications.

Gottman, J.M. & Markman, H.J. (1978). Experimental designs in psychotherapy research. In S.L. Garfield & A.E. Bergin (Eds.), *Handbook of psychotherapy and behavior change*. New York: Wiley.

Green, B.L., Lindy, J.D., Grace, M.C., Gleser, G.C., Leonard, A.C., Korol, M. & Winget, C. (1990). Buffalo Creek survivors in the second decade: stability of stress symptoms. *American Journal of Orthopsychiatry, 60*, 43-54.

Greenson, R.R. and Mintz, T. (1972). California earthquake 1971: Some psy- choanalytic observations. *International Journal of Psychoanalytic Psychotherapy, 1*, 7-23.

Grinker, R.R. & Spiegel, J.P. (1945). *Men under stress*. Philadelphia: Blakiston.

Grübrich-Simitis, I. (1979). Extremtraumatisierung als kumulatives Trauma. *Psyche, 33*, 991-1024.

Gustafson, J.P. (1981). The complex secret of brief psychotherapy in the works of Malan and Balint. In S.H. Budman (Ed.), *Forms of brief therapy*. New York: The Guilford Press.

Haas, J.E. & Drabek, T.E. (1970). Community disaster and system stress: A sociological perspective. In J.E. McGrath (Ed.), *Social and psychological factors in stress* (pp. 264-287). New York: Holt, Rinehart and Winston.

Harel, Z., Kahana, B. & Kahana E. (1988). Psychological well-being among Holocaust survivors and immigrants in Israel. *Journal of Traumatic Stress, 1*, 413-431.

Harshbarger, D. (1976). An ecological perspective on disaster intervention. In H.J. Parad, H.L.P. Resnik & L.G. Parad (Eds.), *Emergency and disaster management* (pp. 271-283).

Helman, C.G. (1990). *Culture, health and illness, Second Edition*. London: Wright.

Helweg-Larsen, P., Hoffmeyer, H., Kieler, F., Hess-Thaysen, E., Thygesen, P. & Wulff, M.H. (1952). Famine disease in German concentration camps. *Acta Psychiatrica Scandinavica*, suppl. *83*.

Helzer, J.E., Robins, L.N., Wish, E. & Hesselbrock, M. (1979). Depression in Vietnam veterans and civilian controls. *American Journal of Psychiatry, 136*, 526-529.

Helzer, J.E., Robins, L.N. & McEvoy, L. (1987). Posttraumatic stress disorder in the general population: Findings of the Epidemiologic Catchment Area survey. *New England Journal of Medicine, 317*, 1630-1634.

Herman, K. & Thygesen, P. (1954). KZ-syndromet. *Ugeskift for Laeger, 116*, 825-836.

Hilgard, J.R. (1970). *Personality and hypnosis: A study of imaginative involvement.* Chicago: The University of Chicago Press.

Hirsch, B.J. (1980). Natural support systems and coping with major life chànges. *American Journal of Community Psychology, 8*, 159-172.

Hobfoll, S.E. & Stephens, M.A.P. (1990). Social support during extreme stress: Consequences and intervention. In B.R. Sarason, I.G. Sarason & G.R. Pierce (Eds.), *Social support: An interactional view* (pp. 454-481). New York: Wiley.

Hofman, M.C., Kleber, R.J. & Brom, D. (1990). *Psychische schade door verkeersongevallen: een signaleringsstudie en een experimenteel preventieproject.* Deventer: Van Loghum Slaterus.

Hollander, A.N.J. den (1970). *Americana.* Meppel: Boom.

Holmes, M.R. & Lawrence, J.S. St. (1983). Treatment of rape-induced trauma: Proposed behavioral conceptualization and review of the literature. *Clinical Psychology Review, 3*, 417-433.

Holmes, T.H. & Rahe, R.H. (1967). The social readjustment rating scale. *Journal of Psychosomatic Research, 11*, 213-218.

Hoppe, K.D. (1971). The aftermath of Nazi persecution reflected in recent psychiatric literature. In N. Krystal & W.G. Niederland (Eds.), *Psychic traumatization: Aftereffects in individuals and communities* (pp. 169-205). Boston: Little, Brown and Company.

Horowitz, M.J. (1976). *Stress response syndromes.* New York: Jason Aronson. (Second Edition was published in 1986).

Horowitz, M.J. (1976b). Diagnosis and treatment of stress response syndromes. In H.J. Parad, H.L.P. Resnik & L.G. Parad (Eds.), *Emergency and disaster management: A mental health sourceback* (pp. 259-269). Bowie, Maryland: Charles Press.

Horowitz, M.J. (1979). Psychological response to serious life events. In V. Hamilton & D.M. Warburton (Eds.), *Human stress and cognition: An information processing approach* (pp. 235-263). Chicester: Wiley.

Horowitz, M. J. & Becker, S. S. (1971). Cognitive responses to stressful stimuli. *Archives of General Psychiatry, 25*, 419-428.

Horowitz, M. J. & Becker, S. S. (1971) The compulsion to repeat trauma: Experimental study of intrusive thinking after stress. *The Journal of Nervous and Mental Disease, 153*, 32-34.

Horowitz, M.J. & Kaltreider, N.B. (1980). Brief psychotherapy of stress response syndromes. In T. Karasu & L. Bellak (Eds.), *Specialized techniques in individual psychotherapy.* New York: Brunner/Mazel.

Horowitz, M. J. & Kaltreider, N.B. (1980). Brief treatment of post-traumatic stress disorders. *New directions for Mental Health Services, 6*, 67-78.

Horowitz, M. J., Marmar, C., Weiss, D. S., DeWitt, K. N. & Rosenbaum, R. (1984). Brief psychotherapy of bereavement reactions: The relationship of process to outcome. *Archives of General Psychiatry, 41*, 438-448.

Horowitz, M.J. & Solomon, G.F. (1975). A prediction of delayed stress response syndromes in Vietnam veterans. *Journal of Social Issues, 31*, 67-80.

Horowitz, M.J., Wilner, N. & Alvarez, W. (1979). Impact of event scale: A measure of subjective stress. *Psychosomatic Medicine, 41*, 209-218.

Hübner, A.H. (1917). Über Kriegs- und Unfallpsychosen. *Archiv für Psychatrie und Nervenkrankheiten, 58,* 324-400.

Hugenholtz, P.Th. (1984). De factoren bij de instandhouding van de psychosociale problematiek van oorlogsgetroffenen. In J. Dane (Red.), *Keerzijde van de bevrijding.* Deventer: Van Loghum Slaterus.

Hull, C.L. (1933). *Hypnosis and suggestibility.* New York: Appleton-Century-Crofts.

Jacobson, E. (1964). *Anxiety and tension control: A physiologic approach.* Philadelphia: Lippincott.

Jager, H. de & Mok, A.L. (1971) *Grondbeginselen der sociologie: gezichtspunten en begrippen.* Leiden: Stenfert.

James, W. (1911). On some mental effects of the earthquake. In *Memories and Studies.* New York: Longmans Green.

Janet, P. (1889). *L'automatisme psychologue.* Paris: Felix Alcan. In 1973 republished - Paris: Societe Pierre Janet.

Janet, P. (1925). *Psychological healing: A historical and clinical study.* New York: The Maxmillan Company. In 1976 republished - New York: Arno Press.

Janis, I. L. (1971). *Stress and frustration.* New York: Harcourt, Brace and World.

Janis, I.L. (1983). Stress inoculation in health care, theory and research. In D. Meichenbaum & M.E. Jaremko (Eds.), *Stress reduction and prevention* (pp. 67-99). New York: Plenum.

Janis, I.L., Chapman, D.W., Gillin, G.M. & Spiegel, J.P. (1955). *The problem of panic.* Washington, D.C.: Federal Civil Defense Administration Bulletin TB-19-2.

Janis, I. L., Defares, P. B. & Grossman, P. (1983). Hypervigilant reactions to threat. In H. Selye (Ed.), *Selye's guide to stress research.* New York: Van Nostrand Reinhold, 1983.

Janney, J.G., Masuda, M. & Holmes, T.H. (1977). Impact of a natural catastrophe on life events. *Journal of Human Stress, 3,* 22-34.

Janoff-Bulman, R. (1989). The benefits of illusions, the threat of disillusionment, and the limitations of inaccuracy. *Journal of Social and Clinical Psychology, 8,* 158-175.

Janoff-Bulman, R. & Frieze, I.H. (1983). A theoretical perspective for understanding reactions to victimization. *Journal of Social Issues, 39,* (2), 1-17.

Jaspars, J., Hewstone, M. & Fincham, F.D. (1983). Attribution theory and research: The state of the art. In J. Jaspars, F.D. Fincham & M. Hewstone (Eds.), *Attribution theory and research: Conceptual, developmental and social dimensions.* London: Academic Press.

Jaspers, J.P.C. (1980). *Gijzelingen in Nederland: een onderzoek naar de psychiatrische, psychologische en andragologische aspecten.* Lisse: Swets & Zeitlinger.

Jessor, R. & Jessor, S.L. (1973). The perceived environment in behavioral science: Some conceptual issues and some illustrative data. *American Behavioral Scientist,* 801-828.

Jones, M.C. (1924). A laboratory study of fear: The case of Peter. *Pedagogical Seminar, 31,* 308-315.

Kalman, G. (1977). On combat-neurosis. *International Journal of Social Psychiatry, 23,* 195-205.

Kardiner, A. (1941). *The traumatic neuroses of war.* New York: Paul Hoeber.

Kazdin, A. E. & Wilcoxon, L. A. (1976). Systematic desensitization and nonspecific treatment effects: A methodological evaluation. *Journal of Abnormal Psychology, 83,* 729-758.

Keane, T.M., Fairbank, J.A., Caddell, J.M., Zimering, R.T. & Bender, M.E. (1985). A behavioral approach to assessing and treating post-traumatic stress disorder in Vietnam veterans. In C. R. Figley (Ed.), *Trauma and its wake: The study and treatment of post-traumatic stress disorder.* New York: Brunner Mazel.

Keane, T.M., & Kaloupek, D.G. (1982). Imaginal flooding in the treatment of a posttraumatic stress disorder. *Journal of Consulting and Clinical Psychology, 50,* 138-140.

Keilson, H. (1979). *Sequentielle Traumatieserung bei Kindern.* Stuttgart: Ferdinand Enke Verlag.

Keiser, L. (1968). *The traumatic neurosis.* Philadelphia: Lippincott.

Kellett, A. (1982). *Combat motivation: The behavior of soldiers in battle.* Boston: Kluwer-Nijhoff.

Kho-So, C. (1977). Naar een diagnose van het gijzelingssyndroom en psychotherapie van gegijzelden. *Tijdschrift voor Psychotherapie, 3,* 206-213.

Kilpatrick, D.G., Resick, P.A. & Veronen, L.J. (1981). Effects of a rape experience: A longitudinal study. *Journal of Social Issues, 37,* 4.

Kilpatrick, D.G., Saunders, B.E., Veronen, L.J., Best, C.L. & Von, J.M. (1987). Criminal victimization: Lifetime prevalence, reporting to police, and psychological impact. *Crime and Delinquency, 33,* 479-489.

Kingsbury, S.J. (1988). Hypnosis in the treatment of posttraumatic stress disorder: An isomorphic intervention. *American Journal of Clinical Hypnosis, 31,* 81-90.

Kinston, W. and Rosser, R. (1974). Disaster: Effects on mental and physical state. *Journal of Psychosomatic Research, 18,* 437-456.

Kinzie, J.D. & Boehnlein, J.K. (1989). Posttraumatic psychosis among Cambodian refugies. *Journal of Traumatic Stress, 2,* 185-199.

Kipper, D. A. (1977). Behavior therapy for fears brought on by war experiences. *Journal of Consulting and Clinical Psychology, 45,* 216-221.

Kleber, R.J. (1982). *Stressbenaderingen in de psychologie.* Deventer: Van Loghum Slaterus.

Kleber, R.J. & Brom, D. (1986). Opvang en nazorg van geweldsslachtoffers in de organisatie. *Gedrag en Gezondheid, 14,* 97-104.

Kleber, R.J. & Brom, D. (1987). Psychotherapy and pathological grief: A controlled outcome study. *Israel Journal of Psychiatry and Related Sciences, 24,* 99-109.

Kleber, R.J. & Brom, D. (1989). Incidentie van posttraumatische stress-stoornissen na frontervaringen, geweldsmisdrijven, ongevallen en rampen. *Tijdschrift voor Psychiatrie, 31,* (10), 675-691.

Kleber, R.J., Brom, D. & Defares, P.B.(1986). *Traumatische ervaringen, gevolgen en verwerking.* [Traumatic experiences: consequences and coping]. Lisse: Swets & Zeitlinger.

Kleinman, A. (1986). *Social origins of distress and disease: Depression, neurasthenia and pain in modern China.* New Haven: Yale University Press.

Kohn, M.L. (1972). Class, family and schizophrenia: A reformulation. *Social Forces, 50,* 295-304.

Kohut, H. (1977). *The restoration of the self.* New York: International Universities Press.

Kolb, L.C. & Brodie, H.K.H. (1982). *Modern Clinical Psychiatry.* Philadelphia: Saunders.

Kolb, L.C. & Mutalipassi, L.R. (1982). The conditioned emotional response: A subclass of the chronic and delayed posttraumatic stress disorders. *Psychiatric Annals, 12*, 979-987.

Kormos, H.R. (1978). The nature of combat stress. In C.R. Figley (Ed.), *Stress disorders among Vietnam veterans* (pp. 3-23). New York: Brunner/Mazel.

Krein, N. (1979). Sudden infant death syndrome: Acute loss and grief reactions. *Clinical Pediatrics, 18*, 414-423.

Kroger, W.C. & Fezler, W.D. (1976). *Hypnosis and behavior modification: Imagery conditioning*. Philadelphia: Lippincott.

Krystal, H. (Ed.) (1968). *Massive psychic trauma*. New York: International Universities Press.

Kulka, R.A., Schlenger, W.A., Fairbank, J.A., Hough, R.L., Jordan, B.K., Marmar, C.R. & Weiss, D.S. (1991). *Trauma and the Vietnam War Veteran generation*. New York: Brunner/Mazel.

Lacey, G.N. (1972). Observations on Aberfan. *Journal of Psychosomatic Research, 16*, 257.

Ladee, G.A. (1975). *Traumatische neurose*. Amsterdam: Winkler Prins Encyclopaedia, Seventh Edition, Part 18, 581.

Landmann, J. & Manis, M. (1983). Social cognition: Some historical and theoretical perspectives. In L. Berkowitz (Ed.), *Advances in experimental social psychology*, volume 16 (pp. 49-126). New York: Academic Press.

Langley, M.K. (1982). Post-traumatic stress disorders among Vietnam combat veterans. *Social Casework - The Journal of Contemporary Social Work, 63*, 593-598.

Lazarus, R.S. (1966). *Psychological stress and the coping process*. New York: McGraw-Hill.

Lazarus, R.S. (1981). The stress and coping paradigm. In C. Eisdorfer, D. Cohen, A. Kleinman & P. Maxim (Eds.), *Models for clinical psychopathology* (pp. 177-214). New York: Spectrum.

Lazarus, R.S. (1983). The costs and benefits of denial. In S. Breznitz (Ed.), *The denial of death* (pp. 1-30). New York: International Universities Press.

Lazarus, R.S. (1984). On the primacy of cognition. *American Psychologist, 39*, 124-129.

Lazarus, R.S. & Launier, R. (1978). Stress-related transactions between person and environment. In L.A. Pervin & M. Lewis (Eds.), *Perspectives in interactional psychology* (pp. 287-327). New York: Plenum.

Lefcourt, H.M. (1980). Locus of control and coping with life's events. In E. Staub (Ed.), *Personality: Basic aspects and current research* (pp. 220-236). Englewood Cliffs, N.J.: Prentice Hall.

Leon, G.R., Butcher, J.N., Kleinman, M., Goldberg, A. & Almagor, M. (1981). Survivors of the holocaust and their children: Current status and adjustment. *Journal of Personality and Social Psychology, 41*, 503-516.

Leopold, R.L. and Dillon, H. (1963). Psycho-anatomy of a disaster: A long term study of post-traumatic neuroses in survivors of a marine explosion. *American Journal of Psychiatry, 111*, 913.

Leymann, H. (1988). Stress reactions after bank robberies: Psychological and psychosomatic patterns. *Work and Stress, 2*, 123-132.

Lick, J. & Bootzin, R. (1975). Expectancy factors in the treatment of fear: Methodological and theoretical issues. *Psychological Bulletin, 82*, 6.

Lifton, R.J. (1979). *The broken connection: on death and the continuity of life*. New York: Simon and Schuster.

Lifton, R.J. (1980). The concept of the survivor. In J.E. Dimsdale (Ed.), *Survivors, victims, and perpetrators: Essays on the Nazi Holocaust* (pp. 113-126). Washington, D.C.: Hemisphere.

Lifton, R.J. & Olson, E. (1976). Death imprint in Buffalo Creek. In H.J. Parad, H.L.P. Resnik & L.G. Parad (Eds.), *Emergency and disaster management* (pp. 295-309).

Lindemann, C. (1944). The symptomatology and management of acute grief. *American Journal of Psychiatry, 101*, 141-149.

Lindy, J. D., Green, B. L., Grace, M. & Titchener, J. (1983). Psychotherapy with survivors of the Beverly Hills supper club fire. *American Journal of Psychotherapy, 4*, 593-610.

Lipkin, J.O., Blank, A.S., Parson, E.R. & Smith, J. (1982). Vietnam veterans and post-traumatic stress disorder. *Hospital and Community Psychiatry, 33*, 908-912.

Lowman, J. (1979). Grief intervention and sudden infant death syndrome. *American Journal of Community Psychology, 7*, 665-677.

Luchterhand, E.G. (1971). Sociological approaches to massive stress in natural and man-made disasters. In H. Krystal & W.G. Niederland (Eds.), *Psychic traumatization: Aftereffects in individuals and communities* (pp. 29-55). Boston: Little, Brown and Company.

Luchterhand, E.G. (1980). Social behavior of concentration camp prisoners: Continuities and discontinuities with pre- and postcamp life. In J. E. Dimsdale (Ed.), *Survivors, victims, and perpetrators: Essays on the Nazi Holocaust* (pp. 259-284). Washington, D.C.: Hemisphere.

Lundberg, V. & Frankenhaeuser, M. (1980). Pituitary-adrenal and sympathetic-adrenal correlates of distress and effort. *Journal of Psychosomatic Research, 24*, 125-130.

Lundh, L.G. (1983). *Mind and meaning: Towards a theory of the human mind considered as a system of meaning structures.* Uppsala, Sweden: Acta Universitatis Upsaliensis, Studia Psychologica Upsaliensis.

Luteijn, F. Starren, J. & Van Dijk, H. (1975). *Handleiding bij de N.P.V.* Lisse: Swets & Zeitlinger.

Maddison, D. (1968). The relevance of conjugal bereavement to preventive psychiatry. *British Journal of Medical Psychology, 41*, 223-233.

Maddison, D.C. & Walker, W.L. (1967). Factors affecting the outcome of conjugal bereavement. *British Journal of Psychiatry, 113*, 1057-1067.

Malan, D. H. (1963). *A study of brief psychotherapy.* New York: Plenum Press.

Malan, D. H. (1976). *The frontier of brief psychotherapy.* New York: Plenum Press.

Malan, D. H. (1979). *Individual psychotherapy and the science of psychodynamics.* London: Butterworths.

Malt, U. (1989). The long-term psychiatric consequences of accidental injury: A longitudinal study of 107 adults. *British Journal of Psychiatry, 153*, 810-818.

Mann, J. (1973). *Time-limited psychotherapy.* Cambridge: Harvard University Press, 1973.

Mann, J. (1981). The core of time-limited psychotherapy: Time and the central issue. In S.H. Budman (Ed.), *Forms of brief therapy.* New York: The Guilford Press.

Manton, M. & Talbot, A. (1990). Crisis intervention after an armed hold-up: Guidelines for counselors. *Journal of Traumatic Stress, 3*, 507-522.

Marmar, C.R. & Freeman, M. (1988). Brief dynamic psychotherapy of posttraumatic stress disorders: management of narcissistic regression. *Journal of Traumatic Stress, 1*, 323-337.

Martin, C.A., McKean, H.E. & Veltkamp, L.J. (1986). Posttraumatic stress disorder in police and working with victims. *Journal of Police Science and Administration, 14,* 98-101.

Matussek, P. (1975). *Internment in concentration camps and its consequences.* New York: Springer Verlag.

Matussek, P. (1981). Holocaust - its roots and consequences. *Contemporary Psychology, 26,* 439-440.

McCaffrey, R.J. & Fairbank, J.A. (1985). Posttraumatic stress disorder associated with transportation accidents: two case studies. *Behavior Therapy, 16,* 406-416.

McCann, I. L. & Pearlman, L.A. (1990). *Psychological trauma and the adult survivor.* New York: Bruner/Mazel.

McFarlane, A.C. (1988). The aetiology of posttraumatic stress disorders following a natural disaster. *British Journal of Psychiatry, 152,* 116-121.

McFarlane, A.C. (1989). The aetiology of posttraumatic morbidity: predisposing, precipitating and perpetuating factors. *British Journal of Psychiatry, 154,* 221-228.

McGee, R.R. & Heffron, E.F. (1976). The role of crisis intervention services in disaster recovery. In H.J. Parad, H.L.P. Resnik & L.G. Parad (Eds.), *Emergency and disaster management* (pp. 309-325).

Meichenbaum, D. (1977). *Cognitive behavior modification: An integrative approach.* New York: Plenum.

Meltzoff, J.E. & Kornreich, M. (1970). *Research in psychotherapy.* New York: Atherton Press.

Meurs, A.J. van der (1955). *Over de gevechtsuitputting.* Rotterdam: Dissertation.

Mitchell, J.T. (1986). Critical incident stress debriefing. *Response!,* 24-25.

Minderhoud, J.M. & Zomeren, A.H. van (1984). *Traumatische hersenletsels.* Utrecht: Bohn, Scheltema & Holkema.

Mirowsky, J. & Ross, C.E. (1989). Psychiatric diagnosis as reified instrument. *Journal of Health and Social Behavior, 30,* 11-25.

Moos, R.H. & Mitchell, R. (1982). Social network resources and adaptation: A conceptual framework. In T.A. Wills (Ed.), *Basis processes in helping relationship.* New York: Academic Press.

Mowrer, O.H. (1960). *Learning theory and behavior.* New York: Wiley.

Mueller, D.P. (1980). Social networks: A promising direction for research on the relationship of the social environment to psychiatric disorder. *Social Science and Medicine, 14A,* 147-161.

Murray, E.J., & Jacobson, L.I. (1978). Cognition and learning in traditional and behavioral therapy. In S.L. Garfield & A.E. Bergin (Eds.), *Handbook of psychotherapy and behavior change, Second Edition.* New York: Wiley.

Nace, E.P., Meyers, A.L., O'Brien, C.P., Ream, N. & Mintz, J. (1977). Depression in veterans two years after Vietnam. *American Journal of Psychiatry, 134,* 167-170.

Nadler, A., Kav-Venaki, S. & Gleitman, R. (1985). Transgenerational effects of the Holocaust: Externalization of agression in second generation Holocaust survivors. *Journal of Consulting and Clinical Psychology, 53,* (3), 365-369.

Nathan, T.S., Eitinger, L. & Winnik, H.Z. (1964). A psychiatric study of survivors of the Nazi holocaust: A study of hospitalized patients. *Israel Annals of Psychology, II,* (1), 47-80.

Neisser, U. (1976). *Cognition and reality: principles and implications of cognitive psychology.* San Francisco: W.H. Freeman.

Niederland, W.G. (1961). The problem of the survivor. *Journal of Hillside Hospital, 10,* 233-247.

Niederland, W.G. (1971). Introductory notes on the concept, definition and range of psychic trauma. In H. Krystal & W.G. Niederland (Eds.), *Psychic traumatization: After-effects in individuals and communities* (pp. 1-11). Boston: Little, Brown and Company.

Novaco, R.W., Cook, T.M. & Sarason, I.G. (1983). Military recruit training: An arena for stress-coping skills. In D. Meichenbaum & M.E. Jaremko (Eds.), *Stress reduction and prevention* (pp. 377-418). New York: Plenum.

Noyes, R., Hoenk, P.R., Kuperman, S. & Slymen, D.J. (1977). Depersonalization in accident victims and psychiatric patients. *Journal of Nervous and Mental Disease, 164,* 401-407.

Nuttin, J. (1968). *Psychoanalyse en persoonlijkheid.* Antwerpen: Standaard.

Ochberg, F.M. (1980). Victims of terrorism. *Journal of Clinical Psychiatry, 41,* (3), 73-74.

Ostwald, P. & Bittner, E. (1968). Life adjustment after severe persecution. *American Journal of Psychiatry, 124,* 1393-1400.

Parad, H.J., Resnik, H.L.P. & Parad, L.G. (Eds.) (1976). *Emergency and disaster management: A mental health sourcebook.* Bowie, Maryland: Charles Press.

Parker, N. (1977). Accident litigants with neurotic symptoms. *Medical Journal of Australia, 2,* 318-322.

Parkes, C.M. (1965). Bereavement and mental illness. *British Journal of Medical Psychology, 38,* 1-26.

Parkes, C.M. (1971). The first year of bereavement: A longitudinal study of the reaction of London widows to the death of their husbands. *Psychiatry Quarterly, 33,* 444.

Parkes, C.M. (1972). *Bereavement: Studies of grief in adult life.* London: Tavistock Publications. (Second Edition was published in 1986).

Parkes, C.M. & Weiss, R.S. (1983). *Recovery from bereavement.* New York: Basic Books.

Parson, E.R. (1986). Life after death: Vietnam veteran's struggle for meaning and recovery. *Death Studies, 10,* 11-26.

Perloff, L.S. (1983). Perceptions of vulnerability to victimization. *Journal of Social Issues, 39,* (2), 41-61.

Peters, J.J. (1975). Social, legal and psychological effects of rape on the victim. *Pennsylvania Medicine, 78,* 34-36.

Peterson, C. & Seligman, M.E.P. (1983). Learned Helplessness and Victimization. *Journal of Social Issues, 39,* (2), 103-116.

Peterson, C. & Seligman, M.E.P. (1984). Causal explanations as a risk factor for depression: theory and evidence. *Psychological Review, 91,* 347-374.

Peterson, K.C., Prout, M.F. & Schwarz, R.A. (1991). *Post-traumatic stress disorder: A clinician's guide.* New York: Plenum Press.

Phares, E.J. (1978). Locus of control. In H. London & J.E. Exner (Eds.), *Dimensions of personality* (pp. 263-305). New York, Wiley.

Pitman, R.K. (1988). Post-traumatic stress disorder, conditioning and network theory. *Psychiatric Annals, 18,* 182-189.

Ploeger, A. (1974). Lengede zehn Jahre danach: medizinisch-psychologische Katamnese einer extremen Belastingssituations. *Psychotherapie und Medische Psychologie, 24,* 137-143.

Powell, J.W. & Rayner, J. (1952). *Progress notes: Disaster investigated* (July 1, 1951-June 30, 1952). Edgewood, Maryland: Army Chemical Center.

Queen's Bench Foundation (1976). *Rape victimization study.* San Francisco: Queen's Bench Foundation, 11-29.
Raapis Dingman, H. (1975). Over de doden niets dan goeds: verslag van een rouwtherapie. *Tijdschrift voor Psychotherapie, 1,* 11-20.
Rabkin, J.G. & Struening, E.L. (1976). Life events stress and illness. *Science, 194,* 1013-1020.
Rachman, S.J. (1978). *Fear and courage.* San Francisco: Freeman.
Rahe, R.H. & Arthur, R.J. (1978). Life change and illness studies: Past history and future directions. *Journal of Human Stress, 4,* 3-15.
Ramsay, R.W. & Happée, J.A. (1977). The stress of bereavement: Components and treatment. In C.D. Spielberger & I.G. Sarason (Eds.), *Stress and anxiety, Volume 4.* Washington, D.C.: Hemisphere.
Ramsay, R.W. (1982). A behavioural approach to anxiety. In H.M. van der Ploeg & P.B. Defares (Red.), *Stress en angst in de medische situatie.* Alphen aan den Rijn: Stafleu, 187-195.
Rangell, L. (1976). Discussion of the Buffalo Creek disaster: The course of psychic trauma. *American Journal of Psychiatry, 133,* 313-316.
Raphael, B. (1977). Preventive intervention with the recently bereaved. *Archives of General Psychiatry, 34,* 1450-1454.
Raphael, B. (1986). *When disaster strikes.* London: Hutchinson.
Rees, W.D. (1971). The hallucinations of widowhood. *British Medical Journal, 4,* 37-41.
Riesman, D. (1950). *The lonely crowd: A study of the changing American character.* New Haven: Yale University Press.
Rogawski, A.S. (1982). Current status of brief psychotherapy. *Bulletin of the Menninger Clinic, 46,* 331-351.
Rosenblatt, P., Walsh, R.P. & Jackson, D.A. (1977). *Grief and mourning in a cross-cultural perspective.* New Haven: Human Relations Area File Press.
Rose, S.L. & Garske, J. (1987). Family environment, adjustment and coping among children of Holocaust survivors: A comparative investigation. *American Journal of Orthopsychiatry, 57,* (3).
Rothbaum, F.M., Weisz, J.R. & Snyder, S.S. (1982). Changing the world and changing the self: A two-process model of perceived control. *Journal of Personality and Social Psychology, 42,* 5-37.
Rotter, J.B. (1966). Generalized expectancies for internal versus external control of reinforcement. *Psychological Monographs, 80.*
Roussy, G. & Lhermitte, J., (1917). *Psychonévroses de guerre.* Paris: Masson et Cie. Editeurs.
Ruch, L.O., Chandler, S.M. & Harter, R.A. (1980). Life change and rape impact. *Journal of Health and Social Behavior, 21,* 248-260.
Russell, A. (1980). Late effects - influence on the children of the concentration camp survivor. In: J.E. Dimsdale (Ed.), *Survivors, victims and perpetrators: Essays on the Nazi Holocaust* (pp. 175-204). Washington D.C.: Hemisphere.
Sanders, C.M. (1988). Risk factors in bereavement outcome. *Journal of Social Issues, 44,* 97-111.
Sandler, I.N. & Lakey, B. (1982). Locus of control as a stress moderator: The role of control perceptions and social support. *American Journal of Community Psychology, 10,* 65-80.

Sarason, I.G. & Sarason, B.R. (Eds.) (1985). *Social support: Theory, research and applications*. Dordrecht: Martinus Nijhoff Publishers.

Sarbin, T.R. & Coe, W.C. (1972). *Hypnosis: A social psychological analysis of the role-taking theory*. New York: Holt, Rinehart & Winston.

Schneider, J.M. (1980). Clinically significant differences between grief, pathological grief, and depression. *Patient Counselling and Health Education*, 161-169.

Seligman, M.E.P. (1975). *Helplessness: On depression, development and death*. San Francisco: W.H. Freeman.

Shaver, K.G. & Drown, D. (1986). On causality, responsibility, and self-blame: A theoretical note. *Journal of Personality and Social Psychology*, *50*, 697-702.

Sherif, M. & Sherif, C.W. (1969). *Social Psychology*. New York: Harper International.

Sifneos, P. E. (1973). The prevalence of 'alexithymic' characteristics in psychosomatic patients. *Psychotherapy and Psychosomatics*, *22*, 255-262.

Sifneos, P.E. (1981) Short-term anxiety-provoking psychotherapy: Its history, technique, outcome and instruction. In S.H. Budman (Ed.), *Forms of brief therapy*. New York: The Guilford Press.

Sigal, J.J., Silver, D., Rakoff, V. & Ellin, B. (1973). Some second generation effects of survival of the Nazi persecution. *American Journal of Orthopsychiatry*, *43*, 320-327.

Silver, R.L., Boon, C. & Stones, M.H. (1983). Searching for meaning in misfortune: Making sense of incest. *Journal of Social Issues*, *39*, (2), 81-102.

Silver, S.M. (1982). Posttraumatic stress disorders in Vietnam veterans: An addendum to Fairbank et al. *Professional Psychology*, *13*, 522-525.

Sims, J.H. & Baumann, D.D. (1972). The tornado threat: coping styles of the North and South. *Science*, *176*, 1386-1392.

Smith, C.A. & Ellsworth, P.C. (1985). Patterns of cognitive appraisal in emotion. *Journal of Personality and Social Psychology*, *48*, 813-838.

Smith, J.R. (1982). Personal responsibility in traumatic stress reactions. *Psychiatric Annals*, *12*, 1021-1030.

Smith, J.R. (1980). *Veterans and combat: Towards a model of the stress recovery process*. Durham, North Carolina: Duke University, internal report.

Smith, M.L., Glass, G.C. & Miller, T.I. (1980). *The benefits of psychotherapy*. Baltimore: Johns Hopkins University Press.

Solomon, Z. & Benbenishty, R. (1986). The role of proximity, immediacy, and expectancy in frontline treatment of combat stress reaction among Israelis in the Lebanon war. *American Journal of Psychiatry*, *143*, 613-617.

Solomon, Z., Weisenberg, M., Schwarzfeld, J. & Mikulincer, M. (1987). Posttraumatic stress disorder among frontline soldiers with combat stress reaction: The 1982 Israeli experience. *American Journal of Psychiatry*, *144*, 448-452.

Sonnenberg, S.M. (1982). Reply to J.I. Walker's 'A disputed diagnosis of posttraumatic stress disorder'. *Hospital and Community Psychiatry*, *33*, 666.

Speisman, J.C., Lazarus, R.S., Mordkoff, A. & Davidson, L. (1964). Experimental reduction of stress based on ego defense theory. *Journal of Abnormal and Social Psychology*, *68*, 367-380.

Spiegel, D. & Cardena, E. (1990). New uses of hypnosis in the treatment of posttraumatic stress disorder. *Journal of Clinical Psychiatry*, *51*, 39-43.

Spielberger, C.D., Gorsuch, R.L. & Lushene, R.E. (1970). *Manual for the state-trait anxiety inventory*. Palo Alto: Consulting Psychologists Press.

Steinmetz, C.H.D. (1984). Coping with a serious crime: Self-help and outside help. *Victimology, 6*, 36-68.

Stierlin, E. (1909). Über psycho-neuropathische Folgezustände bei den Überlebenden der Katastrophe von Courrières am 10. März 1906. *Monatschrift für Psychiatrie und Neurologie, 25*, 185-323.

Stroebe, M.S. & Stroebe, W. (1983). Who suffers more: Sex differences in health risks of the widowed. *Psychological Bulletin, 93*, 279-301.

Stroebe, M.S., Stroebe, W., Gergen, K.J. & Gergen, M. (1981/1982). The broken heart: reality or myth? *Omega, Journal of Death and Dying, 12*, 87-106.

Sutherland, S. & Scherl, D.J. (1970). Patterns of response among victims of rape. *American Journal of Orthopsychiatry, 40*, 503-511.

Swank, R.L. (1949). Combat exhaustion. *Journal of Nervous and Mental Disease, 109*, 475-508.

Swart, J. C. G. de, Grossman, P. & Defares, P.B. (1983). Auditieve ademregulatie ter behandeling van het hyperventilatie syndroom. (Auditive breathing regulation in the treatment of the hyperventilation syndrome). *Tijdschrift voor Geneesmiddelenonderzoek, 8*, 1894-1900.

Symonds, M. (1980). The second injury to victims/acute responses of victims to terror. *Evaluation and Change, special issue*, 36-41.

Tas, J. (1946). Psychische stoornissen in concentratiekampen en bij teruggekeerden. *Maandblad voor Geestelijke Volksgezondheid, 1*, 143-150.

Taylor, S.E. (1983). Adjustment to threatening events: A theory of cognitive adaptation. *American Psychologist, 38*, 1161-1173.

Taylor, S.E., Lichtman, R.R. & Wood, J.V. (1984). Attributions, beliefs about control and adjustment to breast cancer. *Journal of Personality and Social Psychology, 46*, 489-502.

Taylor, V. (1977). Good news about disaster. *Psychology Today, 11*, 93-94, 124-126.

Terr, L. C. (1991). Childhood traumas: An outline and overview. *American Journal of Psychiatry, 148*, 10-20.

Thienes-Hontos, P., Watson, C.G. & Kucala, T. (1982). Stress-disorder symptoms in Vietnam and Korean war veterans. *Journal of Consulting and Clinical Psychology, 50*, 558-561.

Thoits, P.A. (1982). Conceptual, methodological, and theoretical problems in studying social support as a buffer against life stress. *Journal of Health and Social Behavior, 23*, 145-159.

Thompson, S.C. (1981). Will it hurt less if I can control it? A complex answer to a simple question. *Psychological Bulletin, 90*, 89-101.

Thorson, J. (1973). *The long-term effects of traffic accidents*. Stockholm: Karolinska Institutet.

Tirrel, F.J. & Mount, M.K. (1977). A test of a cognitive mediational theory of systematic desensitization. *Journal of Behavior Therapy and Experimental Psychiatry, 8*, 331-332.

Titchener, J.L., Kapp, F.T. & Winget, C. (1976). The Buffalo Creek syndrome: symptoms and character change after a major disaster. In H.J. Parad, H.L.P. Resnik & L.G. Parad (Eds.), *Emergency and disaster management: A mental health sourcebook* (pp. 283-295). Bowie, Maryland: Charles Press. 283-295.

Titchener, J.L. & Ross, W.D. (1974). Acute or chronic stress as determinants of behavior, character, and neurosis. In S. Arieti & E.B. Brody (Eds.), *American Handbook of Psychiatry, Second edition, Volume III* (pp. 39-60). New York: Basic Books.

Titmuss, R.M. (1950). *Problems of social policy.* London: H.M. Stationery Office and Longmans, Green.

Tollison, C.D., Still, J.M. Jr. & Tollison, J.W. (1980). The seriously burned adult: Psychologic reactions, recovery, and management. *Journal of MAG, 69,* 121-124.

Trimble, M.R. (1981). *Post-traumatic neurosis: From railway spine to the whiplash.* Chichester: Wiley.

Tyhurst, J.S. (1951). Individual reactions to community disaster: The natural history of psychiatric phenomena. *American Journal of Psychiatry, 107,* 23-27.

Udolf, R. (1981). *Handbook of hypnosis for professionals.* New York: Van Nostrand Reinhold Company.

Ulman, R.B. & Brothers, (1988). *The shattered self: A psychoanalytic study of trauma.* Hillsdale: The Analytic Press.

Van den Bout, J., Van Son-Schoones, N., Schipper, J. & Groffen, C. (1988). Attributional cognitions, coping behavior, and self-esteem in inpatients with severe spinal cord injuries. *Journal of Clinical Psychology, 44,* 17-22.

Van Dijk, J.J.M. & Steinmetz, C.H.D. (1979). *De WODC-slachtofferenquetes 1974-1979.* The Hague: Staatsuitgeverij.

Van Dijk, W.K. (1981). Gijzelaars en hun familie: een voorbeeld. In J. Bastiaans, D. Mulder, W.K. van Dijk & H.M. van der Ploeg (Red.), *Mensen bij gijzelingen* (pp. 187-214). Alphen aan den Rijn: Sijthoff, 187-219.

Van Dijkhuizen, N. (1980). *From stressors to strains: Research into their interrelationship.* Lisse: Swets & Zeitlinger.

Van Egeren, L. F. (1971). Psychophysiological aspects of systematic desensitization: Some outstanding issues. *Behavioral Research and Therapy, 9,* 65-77.

Van der Hart, O. (1978). *Overgang en bestendiging: over het ontwerpen en voorschrijven van rituelen in de psychotherapie.* Deventer: Van Loghum Slaterus.

Van der Hart, O. (1991) (Ed.). *Trauma, dissociatie en hypnose.* Amsterdam/Lisse: Swets & Zeitlinger.

Van der Hart, O., Brown, P. & Van der Kolk, B.A. (1989) Pierre Janet's treatment of posttraumatic stress. *Journal of Traumatic Stress, 2,* 379-395.

Van der Hart, O. & Friedman, B. (1989) A reader's guide to Pierre Janet on dissociation: a neglected intellectual heritage. *Dissociation, 2,* 3-16.

Van der Kolk, B. A. (1987). *Psychological trauma.* Washington, DC: American Psychiatric Press.

Van der Kolk, B.A. (1988). The trauma spectrum: The interaction of biological and social events in the genesis of the trauma response. *Journal of Traumatic Stress, 1,* 273-290.

Van der Kolk, B.A., Brown, P. & Van der Hart, O. (1989). Pierre Janet on post-traumatic stress. *Journal of Traumatic Stress, 2,* 365-378.

Van der Kolk, B.A. & Van der Hart, O. (1989). Pierre Janet and the breakdown of adaptation in psychological trauma. *American Journal of Psychiatry, 146,* 1530-1540.

Van der Ploeg, H.M. (1980). Validatie van de Zelf Beoordelings-Vragenlijst. *Nederlands Tijdschrift voor de Psychologie, 35,* 243-249.

Van der Ploeg, H.M., Defares, P.B. & Spielberger, C.D. (1981). *Handleiding bij de Zelf-Analyse Vragenlijst Z.A.V.* Lisse: Swets & Zeitlinger.

Van der Ploeg, H.M. & Kleijn, W.C. (1989). Being held hostage in The Netherlands: A study of long-term aftereffects. *Journal of Traumatic Stress, 2*, 153-171.

Van Uden, M.H.F. (1988). Religion in the crisis of bereavement. In E. Chigier (Ed.), *Grief and bereavement in contemporary society, Volume 3* (pp. 91-98). London: Freund Publishing House.

Venzlaff, V. (1964). Mental disorders resulting from racial persecution outside of concentration camps. *International Journal of Social Psychiatry, 10*, 177-183.

Veronen, L.J. & Kilpatrick, D.G. (1983). Stress management for rape victims. In D. Meichenbaum & M. Jaremko (Eds.), *Stress prevention and management: A cognitive behavioral approach* (pp. 341-374). New York: Plenum.

Vingerhoets, A.J.J.M. & Marcelissen, F.G.H. (1988). Stress research: Its present status and issues for future developments. *Social Science and Medicine, 22*, 279-291.

Volkan, V.D. (1970). Typical findings in pathological grief. *Psychiatry Quarterly, 44*, 231-250.

Volkan, V.D. (1979). *Cyprus – war and adaptation: A psychoanalytic history of two ethnic groups in conflicts*. Charlottesville: University of Virginia Press.

Volkart, E.H. & Michael, S.T. (1977). Bereavement and mental health. In S.G. Wilcox & M. Sutton (Eds.), *Understanding death and dying*. Port Washington, New York: Alfred Publishing Co.

Vroom, M.G. (1942). *Schrik, angst en vrees: een psychiatrische en phaenomenologische studie naar aanleiding van vliegtuigbombardementen*. Den Helder, The Netherlands: Uitgeverij de Boer.

Walker, J.I. (1981). Posttraumatic stress disorder after a car accident. *Postgraduate Medicine, 69*, 82-86.

Wallace, A.F.C. (1956). *Tornado in Worchester: An explanatory study of individual and community behavior in an extreme situation*. Washington, D.C.: Committee of Disaster Studies, study no. 3, National Academy of Science.

Watson, J.B. & Rayner, R. (1920). Conditioned emotional reactions. *Journal of Experimental Psychology, 3*, 1-14.

Weisaeth, L. (1989a). A study of behavioural responses to an industrial disaster. *Acta Psychiatrica Scandinavica, 80 Supplementum 355*, 13-24.

Weisaeth, L. (1989b). The stressors and the post-traumatic stress syndrome after an industrial disaster. *Acta Psychiatrica Scandinavica, 80 Supplementum 355*, 25-37.

Weisaeth, L. (1989c). Importance of high response rates in traumatic stress research. *Acta Psychiatrica Scandinavica, 80 Supplementum 355*, 131-137.

Weiss, R.J. & Payson, H.E. (1967). Gross stress reactions. In A.M. Freedman & H.I. Kaplan (Eds.), *Comprehensive Textbook of Psychiatry* (pp. 1027-1031). Baltimore: Williams and Williams.

Weiss, R.S. (1976). The emotional impact of marital separation. *Journal of Social Issues, 32*, 135-145.

Weisz, J.R., Rothbaum, F.M. & Blackburn, T.C. (1984). Standing out and standing in: The psychology of control in America and Japan. *American Psychologist, 39*, 955-969.

WHO (1977). *International statistical classification of diseases, injuries and causes of death, ninth revision (ICD-9)*. Geneva: World Health Organization.

Wientjes, C. J. E., Grossman, P., Gaillard, A. W. K. & Defares, P. B. (1986). Individual differences in respiration and stress. In G. R. J. Hockey, A. W. K. Gaillard & G. H. Coles (Eds.), *Energetics and human information processing*. Dordrecht/ Boston: Martinus Nijhoff, 1986.

Wilde, G.J.S. (1970). *Neurotische labiliteit gemeten volgens de vragenlijstmethode*. Amsterdam: Van Rossem.

Wilkins, K.V. (1981). Unfused homunculi. *The Behavioral and Brain Sciences, 4*, 93-123.

Wilkins, W. (1971). Desensitization: Social and cognitive factors underlying the effectiveness of Wolpe's procedure. *Psychological Bulletin, 76*, 5.

Williams, A.H. (1951). Psychiatric study of Indian soldiers in the Arakan. *British Journal of Medical Psychology, 24*, 130-181.

Williams, R. and Parkes, C.M. (1975). Psychosocial effects of disaster: Birth rate in Aberfan. *British Medical Journal, 2*, 303.

Winnubst, J.A.M., Buunk, A.P. & Marcelissen, F.H.G (1988). Social support and stress: Perspectives and processes. In S. Fisher & J. Reason (Eds.), *Handbook of life stress, cognition and health*. New York: Wiley.

Winnubst, J.A.M., Marcelissen, F.H.G. & Kleber, R.J. (1982). Effects of social supports in the stressor-strain relationship: A Dutch sample. *Social Science and Medicine, 16*, 475-482.

Wolfenstein, M. (1957). *Disaster: A psychological essay*. Glencoe, Ill.: Free Press.

Wolpe, J. (1958). *Psychotherapy by reciprocal inhibition*. Stanford: Stanford University Press.

Worden, J.W. (1982). *Grief counseling and grief therapy: Handbook for the mental health practitioners*. New York: Springer.

Wortman, C.B. (1983). Coping with victimization: Conclusions and implications for future research. *Journal of Social Issues, 39*, (2), 195-221.

Wortman, C.B. & Silver, R.C. (1989). The myths of coping with loss. *Journal of Consulting and Clinical Psychology, 57*, 349-357.

Yamamoto, J., Okonogi, K., Iwasaki, T. & Yoshimora, S. (1969). Mourning in Japan. *American Journal of Psychiatry, 125*, 1660-1665.

Yates, A.J. (1975). *Theory and practice in behavior therapy*. New York: Wiley.

Young, W.C. (1987) Emergence of a multiple personality in a posttraumatic stress disorder of adulthood. *American Journal of Clinical Hypnosis, 29*, 249-254.

Zemore, R. (1975). Systematic desensitization as a method of teaching a general anxiety reducing skill. *Journal of Consulting and Clinical Psychology, 43*, 157-161.

Ziegler, P. (1976). *The black death*. Hammondsworth, Middlesex: Penguin.

Zusman, J. (1976). Meeting mental health needs in a disaster: A public health view. In H.J. Parad, H.L.P. Resnik & L.G. Parad (Eds.), *Emergency and disaster management: A health sourcebook* (pp. 245-259). Bowie, Maryland: Charles Press.

INDEX

I *Names*

Abel, G. 204
Abrahams, M.J. 62, 70, 163
Abramowitz, C.V. 273
Abramowitz, S.I. 273
Abramson, L.Y. 209
Adams, G.R. 62
Adams, P.R. 62
Ahearn, F.L. 54, 57-60
Alexander, F. 127, 233-235, 237
Alvarez, W. 260
Anderson, C.R. 171
Andreasen, N.J.C. 169, 172
Andriessen, J.H.T.H. 262
Anisman, H. 166
Antonovsky, A. 102, 163
Archibald, H.C. 33, 41, 42, 45, 166
Ariès, P. 180, 181
Arindell, W.A. 259, 261
Arthur, R.J. 131
Asher, S.J. 123
Averill, J.R. 107-109, 115, 135, 180, 189, 213

Baker, G.W. 53
Bally, G. 13
Bard, M. 75, 77, 78, 81
Barocas, C. 103
Barocas, H. 103
Barrett, T.W. 176
Barton, A.H. 53, 63
Bastiaans, J. 5, 18, 20, 21, 22, 78, 79, 82, 87, 97, 99, 101, 102, 162-164, 169, 277
Baum, A. 4, 69, 159
Baumann, D.D. 171
Beal, S.M. 123
Becker, S.S. 225
Beebe, B.W. 37, 41, 164, 169
Beecher, H.K. 128
Begemann, F.A. 103
Beigel, A. 52
Benbenishty, R. 40
Bennett, G. 62, 163
Berkman, L.F. 176
Berren, M.R. 52
Best, C.L. 187
Bettelheim, B. 90, 94
Bittner, E. 100, 101
Blackburn, T.C. 147
Blanchard, E.B. 204
Bloom, B.L. 123
Boehnlein, J.K. 183
Boeke, P.E. 161
Boman, B. 167
Boon, C. 145
Bootzin, R. 213
Boudewyns, P.A. 212
Boulanger, G. 280
Bourne, P.G. 38-40, 43, 44
Bowen, G.R. 212
Bowlby, J. 103, 106-109, 112-117, 120-123, 134, 157, 162-164, 167-169, 180, 189, 277
Brandjes, M. 183
Braverman, M. 91
Brende, J.O. 166
Brenner, C. 78, 79
Brett, E.A. 253
Breuer, J. 14-16, 23, 106
Brill, N.Q. 32, 33, 37, 38, 41, 163, 164, 166, 169
Brom, D. 3, 91, 112, 121, 187, 190-192, 200, 216, 258, 261
Brothers, D. 232
Brown, G.W. 166, 176, 177, 212, 249, 253, 277
Buelens, J. 15
Bulman, R.J. 145, 151

Burgess, A.W. 75-77, 79, 81, 82, 84-86, 164, 165, 167, 170
Burnell, A.L. 117, 121
Burnell, G.M. 117, 121
Buunk, A.P. 131

Caddell, J.M. 205
Caplan, G. 191
Card, J.J. 187
Cardena, E. 250, 253
Carey, R.G. 187
Carroll, E.M. 157
Cassee, A.P. 236, 237
Catanese, C.A. 61, 163
Catherall, D.R. 231
Chandler, S.M. 165
Chapman, D.W. 53, 55, 56, 58, 60, 61
Charmaz, K. 180
Chertok, L. 244, 245
Chodoff, P. 99
Clayton, P.J. 175
Cleary, P.D. 70
Coe, W.C. 245, 257
Cohen, E.A. 94-96, 158
Cohen, R.E. 54, 57-60
Cook, T.M. 198
Culpan, R. 13

Danieli, Y. 103
Dasberg, H. 102
Davidson, J. 258, 280
Davidson, L.M. 69 ,132
Davidson, S. 177
Defares, P.B. 3, 80, 170, 183, 216, 219, 220, 258, 260, 261
DeFazio, V.J. 38, 41, 43, 178
De Frain, J.D. 189
De Graaf, Th. 104
Dennett, D.C. 143
De Wind, E. 96
De Wolf, J. 236
Dillon, H. 67, 163
Dimsdale, J.E. 94
Dohrenwend, B.P. 127, 164
Dohrenwend, B.S. 127, 164
Donahoe, C.P. 157
Donovan, D.M. 104
Dor-Shav, N.K. 102
Doreleijers, T.A.H. 104
Dowty, N. 102
Drabek, T.E. 53
Duffy, B. 81
Duyker, H.C.J. 9
Dynes, R.R. 54, 56, 61, 63

Eissler, K. 105
Eitinger, L. 96, 97, 99-101
Eland, J. 104
Ellemers, J.E. 54, 57
Ellin, B. 102
Ellsworth, P.C. 135
Endler, N.S. 49, 173
Engel, G.L. 189
Erichsen, J.E. 12, 13
Erickson, M.H. 245, 254
Erikson, K.T. 64, 65, 179
Ernst, L. 189
Eth, S. 7
Ettema, H. 259, 261
Evans, H.I. 176
Evans, I.M. 213
Ewalt, J.R. 45, 179

Fairbank, J.A. 43, 45, 205, 212
Fenichel, O. 17, 18, 61
Ferenczi, S. 234
Fezler, W.D. 243, 248
Figley, C.R. 45
Fincham, F.D. 88
Finkel, N.J. 33
Flannery, R.B. 176, 178
Fleming, R. 69
Foeckler, M.M. 91, 165, 169, 176, 187
Folkman, S. 135
Foy, D.W. 157, 176
Frank, E. 81, 82
Frankl, V.E. 99, 147, 282
Frederick, C.J. 56, 58, 61, 88
Freeman, M. 231, 240
French, J.R.P. 131
French, T.M. 233, 234, 237

Freud, S. 4, 5, 12, 14-17, 23, 32, 33, 106, 110, 114, 115, 121, 137, 138, 140, 145, 152, 189, 224, 226, 228, 233, 234, 244
Freyberg, J.T. 102, 103
Friedman, P. 57, 59, 60, 96, 249
Friedsam, H.J. 61, 163
Frieze, I.H. 144, 278
Frijda, N.H. 108, 134, 135
Fritz, C.E. 58, 59
Fromm, E. 244, 245, 250, 251
Fromm-Reichmann, F. 187
Frye, J.S. 45, 46, 48, 171, 179
Furst, S. 16

Gaillard, A.W.K. 219
Garfield, S.L. 262
Garrard, F.H. 91, 187
Garske, J. 103
Gergen, K.J. 128
Gergen, M. 128
Gersons, B.P.R. 199
Ghertner, S. 52
Giel, R. 69
Gilligan, S.G. 254
Gillin, G.M. 58
Glass, A.J. 39
Glass, G.C. 259
Gleitman, R. 103
Glick, I.O. 107, 113, 158, 177
Goldberg, A. 103, 231
Goldfried, M.R. 214, 215
Goldney, R.D. 166
Gorer, G. 113, 123, 181
Gorsuch, R.L. 261
Gottman, J.M. 212
Grace, M.C. 68, 258
Green, B.L. 68, 258
Greenson, R.R. 54
Grinker, R.R. 17, 18, 35-39
Groffen, C. 146
Grossman, P. 219, 220
Grübrich-Simitis, I. 99
Gustafson, J.P. 235, 237

Haas, J.E. 53
Harel, Z. 102, 177
Harris, T. 177
Harrison, W.R. 212
Harshbarger, D. 63
Harter, R.A. 165
Hartford, C.R. 169
Heffron, E.F. 62
Helman, C.G. 179, 181
Helweg-Larsen, P. 96
Helzer, J.E. 44, 45, 163
Herman, K. 97
Hewstone, M. 88
Hilgard, J.R. 245, 249
Hirsch, B.J. 176, 177
Hobfoll, S.E. 178
Hoenk, P.R. 91
Hoffmeyer, H. 96
Hofman, M.C. 91, 92, 192
Hollander, A.N.J. den 65
Holmes, M.R. 78, 79, 80, 212
Holmes, T.H. 4, 22, 54, 127, 165
Hoppe, K.D. 99
Horowitz, M.J. 9-11, 20, 22-25, 44, 79, 88, 137-143, 151, 188, 223-229, 231, 238, 239, 241, 258, 260, 261, 272, 277
Hough, R.L. 45
Houts, P.S. 70
Hübner, A.H. 33
Hugenholtz, P.Th. 95
Hull, C.L. 245
Hurt, S.W. 251
Hyer, L. 212

Iwasaki, T. 180

Jackson, C. 273
Jackson, D.A. 180
Jacobson, E. 212, 216, 219
Jager, H. de 259
James, W. 53
Janet, P. 14, 23, 248, 249
Janis, I.L. 22, 58, 73, 121, 137, 140, 158, 159, 219, 227
Janney, J.G. 54
Janoff-Bulman, R. 73, 124, 144, 150, 151, 278
Jarvie, G.J. 43

Jaspars, J. 88
Jaspers, J.P.C. 169
Jessor, R. 165
Jessor, S.L. 165
Jones, M.C. 211
Jones, T.J. 91, 187
Jordan, B.K. 45

Kahana, B. 102
Kahana, E. 102
Kahn, R.L. 131
Kalman, G. 36, 166
Kaloupek, D.G. 212
Kaltreider, N.B. 22, 238, 239
Kapp, F.T. 60
Kardiner, A. 17
Kav-Venaki, S. 103
Kazdin, A.E. 213
Keane, T.M. 43, 205, 212, 220
Keilson, H. 99, 179
Keiser, L. 13, 16
Kellett, A. 32, 38
Kho-So, C. 5
Kieler, F. 97
Kilpatrick, D.G. 76, 82, 84-87, 187, 198, 212
Kingsbury, S.J. 250
Kinston, W. 54-56, 60
Kinzie, J.D. 183
Kipper, D.A. 204
Kleber, R.J. 3, 6, 8, 19, 21, 91, 104, 112, 121, 128, 150, 153, 176, 187, 190-192, 199, 216, 258, 261, 278
Kleijn, W.C. 79, 84, 85, 196
Kleinman, A. 182
Kleinman, M. 103
Kohn, M.L. 164
Kohut, H. 168, 231, 232
Kolb, L.C. 205
Kormos, H.R. 33-35, 39, 41
Kornreich, M. 264
Krein, N. 123
Kroger, W.C. 243, 248
Krystal, H. 99
Kucala, T. 48
Kulka, R.A. 45, 46, 157, 277
Kuperman, S. 91

Ladee, G.A. 20
Lakey, B. 171
Lambert, J.A. 212
Landmann, J. 137
Langley, M.K. 43
Launier, R. 133
Lawrence, J.S.St. 79, 80, 212
Lazarus, R.S. 8, 19, 22, 126, 132, 133-135, 137, 140, 144, 153, 154, 158, 161, 172, 177, 183, 227, 283
Lefcourt, H.M. 170, 172
Leon, G.R. 103
Leonard, A.C. 68
Leopold, R.L. 67, 163, 164
Leymann, H. 163
Lhermitte, J. 33
Lichtman, R.R. 146
Lick, J. 213
Lifton, R.J. 25, 64, 65, 94, 98, 99, 105, 108, 147
Lindemann, C. 106, 108
Lindy, J.D. 68, 258
Linn, L. 57, 59, 60
Lipkin, J.O. 28, 45
Long, D.M. 41
Lowman, J. 189
Luchterhand, E.G. 88, 89, 96
Lundh, L.G. 134
Lushene, R.E. 261
Luteijn, F. 261

Maddison, D.C. 117, 168, 175
Magnusson, D. 49, 173
Malan, D.H. 234-237, 241
Malt, U. 91
Manis, M. 137
Mann, J. 237, 241
Maoz, B. 102
Marcelissen, F.H.G. 130, 131, 176
Markman, H.J. 212
Marks, E.A. 59
Marmar, C.R. 45, 231, 240, 258
Martin, C.A. 163
Masuda, M. 54

Matussek, P. 97, 100
McCaffrey, R.J. 212
McCann, I.L. 155, 205, 233, 278
McEvoy, L. 163
McFarlane, A.C. 54, 157, 169
McGee, R.R. 62
McKean, H.E. 163
Meichenbaum, D. 198
Meltzoff, J.E. 264
Meurs, A.J. van der 39
Meyers, A.L. 44
Michael, S.T. 180
Miller, C. 41
Miller, T.I. 259
Minderhoud, J.M. 92
Mintz, J. 44
Mintz, T. 54
Mirowsky, J. 29
Mitchell, J.T. 71, 197, 198
Mitchell, R. 176
Mizes, J.S. 176
Mok, A.L. 259
Moos, R.H. 176
Mordkoff, A. 132
Mount, M.K. 214
Mowrer, O.H. 205, 247
Mueller, D.P. 176
Murray, E.J. 212
Mutalipassi, L.R. 205

Nace, E.P. 44, 166
Nadler, A. 103
Nass, C.H.Th. 183
Neisser, U. 134, 136
Niederland, W.G. 97, 99, 101
Norris, A.S. 169
Novaco, R.W. 198
Noyes, R. 91, 277
Nuttin, J. 15

O'Brien, C.P. 44
Ochberg, F.M. 91
Okonogi, K. 180
Olson, E. 64, 65
Ostroff, R. 253
Ostwald, P. 100, 101

Parkes, C.M. 63, 106-113, 117-120, 123, 124, 158, 162, 163, 165, 168, 169, 173, 175-177, 181
Parson, E.R. 28, 233
Payson, H.E. 55
Pearlman, L.A. 155, 205, 233, 278
Perloff, L.S. 150, 151
Peters, J.J. 79, 84
Peterson, C. 144, 148
Peterson, K.C. 28, 87, 179, 204
Petrie, J.F. 236
Phares, E.J. 171
Pitman, R.K. 208
Ploeger, A. 54
Powell, J.W. 54, 55
Price, J. 62
Prout, M.F. 28, 87
Pynoos, R.S. 7

Quarantelli, E.L. 54, 56, 58, 61, 63
Queen's Bench Foundation 81, 84

Raapis Dingman, H. 120
Rabkin, J.G. 128, 166
Rachman, S.J. 37-39, 159, 160, 164, 174, 175
Rahe, R.H. 4, 22, 127, 131, 165
Rakoff, V. 102
Ramsay, R.W. 80, 108, 114
Rangell, L. 63
Rank, O. 234
Raphael, B. 54, 71, 192
Rayner, J. 54, 55
Rayner, R. 211
Ream, N. 44
Rees, W.D. 110, 113
Resick, P.A. 212
Riesman, D. 174
Roback, H.B. 273
Robins, L.N. 44, 163
Rogawski, A.S. 236
Rose, S.L. 103
Rosenblatt, P. 180
Ross, C.E. 19
Ross, W.D. 29
Rosser, R. 54-56, 60

Rothbaum, F.M. 144, 147, 152, 279
Rotter, J.B. 170
Roussy, G. 33
Ruch, L.O. 165
Russell, A. 103

Sanders, C.M. 162, 164
Sandler, I.N. 171
Sangrey, D. 75, 77, 78, 81
Sarason, B.R. 176
Sarason, I.G. 176, 198
Sarbin, T.R. 245, 257
Saunders, B.E. 187
Scherl, D.J. 75, 77, 78, 83, 86, 176
Schipper, J. 146
Schlenger, W.A. 45
Schneider, J.M. 120, 121
Schwarz, R.A. 28, 87
Seligman, M.E.P. 144, 147, 148, 150-152, 208, 209, 215
Shaver, K.G. 146
Sherif, C.W. 128, 175
Sherif, M. 128, 175
Sifneos, P.E. 235-237
Sigal, J.J. 102, 103
Silver, D. 102
Silver, R.C. 114, 116, 117
Silver, R.L. 145
Silver, S.M. 43
Sims, J.H. 171
Singer, J.E. 4, 69, 159
Slymen, D.J. 91
Smith, C.A. 135
Smith, J. 28
Smith, J.R. 160, 184
Smith, M.L. 259
Smith, R. 258, 280
Snyder, S.S. 144, 279
Solomon, Z. 34, 40, 44
Sonnenberg, S.M. 124
Speisman, J.C. 132
Spiegel, D. 250, 253
Spiegel, J.P. 17, 18, 35-39, 58, 174
Spielberger, C.D. 261
Starren, J. 261
Steinmetz, C.H.D. 73, 90, 104
Stephens, M.A.P. 178
Stierlin, E. 53, 54
Still, J.M. Jr. 91
Stockton, R.A. 45, 46, 48, 171, 179
Stones, M.H. 145
Stroebe, M.S. 128, 158, 163, 164
Stroebe, W. 128, 163, 164
Strøm, A. 101
Struening, E.L. 128, 166
Sutherland, S. 75, 77, 78, 83, 86, 176
Swank, R.L. 34-38, 40, 169
Swart, J.C.G. de 220
Syme, S.L. 176
Symonds, M. 75, 76, 81

Tas, J. 96
Taylor, C. 13
Taylor, S.E. 54, 144-146, 149, 150-152, 282
Taylor, V. 54, 61, 66, 192
Teasdale, J.D. 209
Terr, L.C. 276
Thienes-Hontos, P. 47, 48
Thoits, P.A. 165, 17
Thomas, A.M. 91, 187
Thompson, S.C. 150, 189
Thorson, J. 91
Thygesen, P. 97
Tirrel, F.J. 214
Titchener, J.L. 19, 59, 64, 68
Titmuss, R.M. 53
Tollison, C.D. 91
Tollison, J.W. 91
Trimble, M.R. 13, 32
Tuddenham, R.D. 33, 41, 42, 166
Turner, S.M. 81
Tyhurst, J.S. 53, 55, 57, 59

Udolf, R. 255
Ulman, R.B. 232

Van den Berg, J.F. 79
Van den Berg-Schaap, Th.E. 79
Van den Bout, J. 146
Van Dijk, H. 261
Van Dijk, J.J.M. 73

Van Dijk, W.K. 73, 76, 77, 82, 83, 90
Van Dijkhuizen, N. 131
Van Egeren, L.F. 213
Van der Hart, O. 11, 12, 14, 249, 253, 260, 277
Van der Kolk, B.A. 14, 154, 196, 233, 249, 253, 277
Van der Ploeg, H.M. 79, 84, 85, 196, 261
Van der Ploeg, J.D. 183
Van Son-Schoones, N. 146
Van Uden, M.H.F. 112, 168, 181
Van der Velden, P.G. 104
Veltkamp, L.J. 163
Venzlaff, V. 97, 100
Veronen, L.J. 76, 82, 84-87, 187, 198, 212
Vingerhoets, A.J.J.M. 130
Volkan, V.D. 18, 120, 122, 189
Volkart, E.H. 180
Von, J.M. 187
Vroom, M.G. 53

Walker, J.I. 91, 117, 175
Wallace, A.F.C. 60
Walsh, R.P. 180
Watson, C.G. 48
Watson, J.B. 211
Weisaeth, L. 54, 58, 86, 157, 196, 198
Weiss, D.S. 45, 258
Weiss, R.J. 55
Weiss, R.S. 107, 108, 113, 117-119, 123, 162, 168, 181
Weisz, J.R. 144, 147, 279
White, S.W. 123
Whitlock, F.A. 62
WHO (1977) 19
Wientjes, C.J.E. 219
Wijsenbeek, H. 102
Wilcoxon, L.A. 213
Wilde, G.J.S. 262
Wilkins, K.V. 213
Wilkins, W. 143
Williams, A.H. 40, 182
Williams, C.C. 91, 187
Williams, G. 62
Williams, R. 63
Wilner, N. 260
Winget, W. 60
Winnubst, J.A.M. 131, 176
Wish, E. 44
Wolfenstein, M. 55, 57, 58, 60, 158, 167, 189
Wolpe, J. 211, 212, 214, 215
Wood, J.V. 146
Woods, M.G. 212
Worden, J.W. 114, 115
Wortman, C.B. 88, 114, 116, 117, 145, 151

Yamamoto, J. 180
Yates, A.J. 213
Yoshimora, S. 180
Young, W.C. 249

Zacharko, R.M. 166
Zemore, R. 214
Ziegler, P. 51
Zusman, J. 60, 62

INDEX

II *Subjects*

Accident 3, 4, 8, 13, 17, 18, 26, 69, 70, 90-92, 145, 150, 163, 169, 190, 199, 202, 209, 278
Accident, traffic 2, 11, 91, 92, 144, 165, 176, 189, 192, 193, 199, 209, 216, 259, 278, 279
Acute stress reaction 19, 20, 26, 29
Adjustment reaction/disorder 19, 26, 28, 29, 30
Age, as a determinant 38, 42, 62, 67, 68, 85, 87, 129, 161-163, 166, 259, 270-272
Aggression 17, 36, 57, 80, 95, 99, 101, 104, 105, 166, 183
Amnesia 14, 23, 27, 35, 65, 92, 249
Anger 14, 26, 60, 61, 64, 65, 76, 77, 81, 83, 90, 93, 95, 98, 99, 109, 111, 112, 114, 116, 120, 123, 127, 134, 140, 158, 168, 175, 180, 183, 214, 217-221, 240, 248, 249, 260, 263, 270, 271, 273, 277
Anticipation 55, 60, 91, 122, 129, 133, 158, 159, 163
Anxiety 13, 14, 16, 18, 25, 29, 33-36, 38, 39, 41, 42, 53, 66, 70, 76, 78-82, 84, 90, 93, 117-119, 131, 133, 140, 147, 151, 154, 158, 159, 162, 163, 166, 170, 175, 177, 183, 206, 207, 211, 212, 214-218, 221, 224, 227, 235, 237, 240, 247, 260, 263, 266-268, 279
Appraisal 132-137, 153, 154, 159, 175, 183, 191, 194, 210, 227, 228, 236
Assumptions, basic 73, 137, 138, 142, 143, 151, 155, 189, 190, 278
Attachment theory 103, 107, 113, 122, 134, 167, 168, 180
Attribution(s) 146, 151, 153, 209, 210, 215
Attribution of meaning 128-132, 134, 135, 152-154, 168, 257, 280, 282, 283
Attribution theory 83, 88, 144, 145, 148, 149, 209
Avoidance 8, 22, 24-26, 46, 54, 60, 69, 74, 75, 77, 82, 86, 88, 92, 99, 110, 111, 115, 117, 120, 154, 158, 188, 195, 200, 203, 207, 208, 215, 221, 239, 248, 254, 256, 261, 263-265, 269, 272, 281

Bank robbery 198, 199, 203
Behavior theory 204, 222, 243
Behavior therapy 10, 204, 205, 211, 212, 248, 257
Bereavement (*see also* Loss) 107, 108, 112, 114-117, 119, 120, 122-124, 163, 164, 169, 180, 181
Bewilderment 23, 57-59, 74, 75, 92, 108, 109, 142, 155, 158, 192
Biographical characteristics, *see*:
- Age
- Education
- Gender
- Social class

Breathing 216, 219, 220, 245
Buffalo Creek 53, 63, 65, 66, 68, 157

Cambodia 183
Causality, attribution of 148, 149, 170, 209, 210
Central conflict 236, 237
Chernobyl 52, 69
Childhood experiences, as a determinant 9, 42, 166-169, 171
Classical conditioning 10, 80, 205, 206, 214, 221
Combat exhaustion 20, 33-41, 43, 46, 48, 49, 163-165, 169, 182
Combat stress 8, 32, 34, 40, 71, 183, 186
Compulsive repetition, *see* Repetition compulsion
Concentration, disturbances in 23, 26, 35, 80, 195, 261

Concentration camp 8, 18, 64, 90, 94, 96-100, 102, 104, 105, 127, 158, 163, 173
Concentration camp syndrome 5, 94, 97, 99, 104, 105
Confrontation, as a technique 192, 194, 195, 197, 206-208, 211, 212, 214, 220, 221, 247, 248, 253, 256, 272, 281
Control, concept of 5, 149-153, 155
Control, interpretational/cognitive 147, 149, 150, 152, 189, 239
Control processes 139-145, 147, 149-153, 155
Coping, concept of 23, 30, 126, 153
Counter-conditioning 212, 214
Crimes of violence 3, 8, 72-78, 81-90, 92, 93, 138, 150, 153, 162, 199, 201, 203, 210, 228, 259, 277-279
Crisis intervention 191, 192
Culture and traumatic stress 65, 68, 71, 122, 174, 178-180, 182, 184, 202, 203

Death imprint 25, 64, 98, 147
Debriefing 197
Defense 22, 65, 77-79, 86, 98, 110, 115, 166, 209, 232, 237, 251, 276
Defense mechanism 54, 65, 78, 79, 138, 142, 226, 239, 249, 253, 277
Denial 22-24, 28, 54-56, 65, 71, 78, 79, 86, 96, 98, 101, 110, 123, 139, 140, 142, 153, 161, 187, 188, 192, 195, 225, 227, 228, 230, 239
Depersonalization 75, 76, 95, 133, 277
Depression 27, 28, 36, 44, 45, 53, 64-67, 71, 80, 82, 92, 103, 112, 115-121, 124, 127, 148, 151, 166, 167, 172, 175, 177, 182, 183, 190, 209, 279
Desensitization, in general 212-216, 218-222, 248, 253
Desensitization, systematic 204, 207, 211-213
Desensitization, trauma 204, 211, 215-217, 219-221, 260, 264-266, 270-272, 280-282

Determinants of coping, *see*:
- Age
- Anticipation
- Childhood experiences
- Culture
- Education
- Gender
- Locus of control
- Marital status
- Personality
- Personality characteristics
- Social class
- Social support
- Society and traumatic stress
- Socio-demographic background
- Stressfull life events

Disaster 3, 8, 17, 18, 23, 26, 51-68, 70, 71, 75, 88, 89, 98, 127, 129, 150, 153, 157-159, 162, 169, 171, 178, 179, 186, 193, 197, 198, 278
Disaster, man-made 8, 52-54, 62, 71, 86
Disaster, natural 3, 4, 7, 8, 19, 52, 53, 71, 88, 89, 129, 171
Disaster, technological/ecological 52, 69, 70
Disaster syndrome 60
Disbelief 23, 74-76, 88, 91-93, 95, 108, 118
Disorganization, phase of 108, 111, 112
Disruption 4-6, 21, 49, 51, 60, 73, 84, 91, 95, 98, 123, 142, 144, 155-157, 187, 188
Dissociation 14, 190, 245, 248-250, 253, 257, 277
Divorce 6, 44, 123, 145, 259
DSM-I 19, 33
DSM-II 25
DSM-III 25, 26, 28, 29
DSM-III-R 25-27, 29

Earthquake 8, 51-54, 59
Education, as a determinant 38, 42, 45, 67, 68, 85, 87, 161, 164, 259
Exposure, as a technique 208, 212, 239, 253

Flashbacks 69, 208
Flood 8, 51-54, 57, 60-63, 65, 66, 68, 71
Flooding, as a technique 204, 207, 212
Focus 236, 237, 241, 242
Frozen fright 76

Gender, as a determinant 46, 47, 62, 68, 85, 161, 163, 259
Grief 63, 107-109, 111, 112, 114, 115, 117, 120, 123, 124, 140, 141, 157, 162, 180, 181, 182, 189, 277
Grief, pathological 107, 115-117, 119-122, 158, 162, 163, 168, 169, 175, 181
- Grief, delayed 117, 121
- Grief, unexpected/unanticipated 117, 118
- Grief, conflicted 117, 118, 119
- Grief, chronic 116, 117, 119, 121
Gross stress reactions 19, 33, 49
Group assistance 196, 197
Guilt 25, 46, 49, 63-65, 77, 81-83, 88, 89, 91, 93, 98, 103, 104, 118, 119, 123, 133, 202, 217-219, 221, 277, 279

Hallucinations 27, 60, 110, 180
Helplessness 5, 6, 14, 16, 58, 62, 75, 81, 82, 90, 95, 119, 121, 148-150, 198, 205, 209, 249
Heroic phase 56
Hierarchy of stimuli 211, 212, 214, 216, 218-220
Hijacking 5, 8, 10, 73, 76-79, 82, 83, 85, 87, 90, 91, 162, 164, 169, 186, 199
Hiroshima 25, 64, 98
Hold-up 199, 201, 205
Holocaust 102, 103, 179, 231
Honeymoon phase 56, 59-61, 89
Hostage 2, 5, 72, 73, 76-80, 82-85, 87, 90, 91, 162, 164, 169, 170, 202
Hostility (*see also* Anger) 61, 77, 88, 261, 266, 270
Hyperalertness 26, 42, 102, 105, 148
Hyperventilation 253
Hypno-behavioral model 247, 248
Hypnosis 11, 14, 40, 233, 243-246, 248-255, 257, 281
Hypnotherapy 11, 243, 250, 254, 257, 260, 264-266, 270-272, 281, 282
Hysteria 13-16, 18, 233

ICD 19, 20, 29
Identification with the aggressor 90, 91, 93, 95
Illness 59, 60, 62, 69, 96, 97, 100, 102, 126-128, 130, 131, 146, 150, 157, 161, 165, 172, 176, 182, 188, 259, 278
Illusion 57, 58, 89, 109, 110, 138, 150, 151, 189, 190, 228, 278, 279
Imagery 219, 221, 226, 246, 248, 250, 253, 256, 272, 277, 278, 281
Immediacy, proximity and expectancy 39, 40, 200
Incidence and prevalence of PTSD 45-48, 84, 85, 102, 121, 187
Information processing 11, 142, 223, 224, 228, 231, 240, 250
Interpersonal problems, *see* Social inadequacy
Intrusion 22-24, 27, 28, 60, 71, 72, 75, 79, 81, 92, 98, 100, 101, 103, 139, 140, 142, 187, 188, 192, 195, 217, 225, 230, 261, 263-265, 269, 272
Invulnerability 55, 58, 73, 138, 147, 150, 151, 189, 201, 228, 278, 279

KZ syndrome, *see also* Concentration camp syndrome 97

Latency phase/outward adjustment 77, 78, 101, 192
Learned helplessness 147, 148, 152, 160, 164, 208-210, 215
Learning theory 80, 81, 140, 144, 148, 154, 170, 204, 205, 207, 208, 245, 247, 248, 281
Locus of control 170-172, 191, 270, 271
Loss of a loved one 8, 107, 108, 113, 123, 124, 181, 259, 277, 278

Marital status, as a determinant 38, 68, 112, 118, 119, 161, 164, 165, 168, 180
Memory, disturbances in 36, 60, 195, 261

Methodological issues 85, 87, 173
Michigan model 130
Mood disorders, *see also* Depression 27, 30, 82
Mourning, *see also* Grief 18, 63, 91, 106-108, 111, 113-116, 119-122, 146, 180, 181
Multiple personality 27, 249
Muselman behavior 96

Narcissistic injury 231, 232, 240-242, 251
Neuroticism 169, 170
Nightmare 22, 23, 35, 38, 45, 46, 49, 60, 64, 84, 99, 100, 145, 188, 211
Numbness 22-25, 57, 59, 63, 65, 71, 75, 77, 98, 104, 108, 109, 116, 148, 188, 279

Operant conditioning 80, 207, 214
Organization, traumatic stress in the 186, 196, 198-203, 280
Organization, intervention in the 199-202
Outcry 12, 23, 24, 225

Panic 58
Personality 129-131, 166, 168, 210, 222, 267, 270
Personality characteristics 9, 20, 21, 30, 36, 37, 42, 85, 87, 122, 127, 160, 161, 166-170, 173, 176, 191, 207, 219, 221, 254, 261-263, 272
Personality disorders 19, 43, 231, 232
Personality, changes in 20, 29, 36, 100, 105, 173, 237, 238, 272
Phase models 23, 24, 54-56, 74, 75, 101, 102, 108, 114, 225
Phobia 7, 64, 80, 81, 84, 99, 261, 266
Physiologic reactivity 26
Positive reactions after trauma 63, 66, 82
Post-disaster utopia 89
Posttraumatic stress disorder (PTSD) 3, 10, 12, 25-30, 43, 46-49, 92, 124, 157, 163, 169, 170, 172, 182, 183, 190, 191, 197, 202, 204, 208, 217, 222, 223, 231, 238, 249-251, 253, 254, 258, 259, 270, 271
Psychoanalysis 9, 10, 12, 16, 17, 22, 30, 36, 107, 137, 139, 235, 250
Psychodynamic therapy 223, 224, 233-235, 241, 250, 260, 264-266, 268, 270-272, 281, 282
Psychodynamic theory 154, 239, 240, 249-251
Psychotic reactions to trauma 19, 27, 28, 30, 42, 61, 182, 183, 190

Rage 18, 65, 100, 103, 240, 252, 253
Railway spine 13
Rape 8, 72, 75, 76, 79, 81-84, 89, 90, 165, 198, 212
Relaxation 40, 43, 44, 211, 212, 214-217, 219-221, 223, 245, 252, 254-256, 281
Reorganization/integration, phase of 74, 75, 83-85, 108, 112, 120
Repetition compulsion 138, 228, 229
Replica-dream 81
Risk factors, *see also* Determinants of coping 196, 198, 200, 203, 283
Rituals 108, 122, 174
Robbery 8, 72, 73, 89, 186, 201

Schemata 135-144, 146, 155, 162, 184, 189, 228, 230, 240, 277
Search for control 152, 153
Search for meaning 81, 83, 99, 144, 146, 147, 152, 153, 189, 276
Secondary victimization 90
Self-blame 88, 93, 116, 121, 145, 146, 279
Self-help groups 198, 199
Self-image 121, 138, 151, 238, 240, 278
Self-psychology 196, 223, 231, 240, 250
Shell shock 16, 32
Sleep disturbances 18, 35, 42, 53, 59, 71, 80, 82, 84, 85, 100, 110, 116, 211, 261
Social class, as a determinant 42, 161, 164, 259
Social comparison 56, 63, 144, 150, 151

Social inadequacy 45, 47, 50, 60, 65, 66, 82, 84, 93, 100, 101, 118, 119, 182, 261, 266
Social support 9, 44, 59, 61, 63, 71, 89, 112, 113, 122, 130, 142, 145, 164, 166, 167, 174-179, 181, 183, 184, 191, 193, 194
Society and traumatic stress 48, 64, 65, 74, 89, 90, 96, 99, 113, 122, 174, 178, 179, 199
Sociodemographic background 161-165, 262
Somatic symptoms 14, 59, 60, 65, 69, 71, 84, 182, 260, 261, 266
Startle response/reactions 22, 23, 26, 36, 45, 69, 99, 205
Stimulus barrier 16, 140, 224, 226
Stress, as a concept 21, 29
Stress inoculation 159, 198
Stress response syndrome 7, 22, 24, 25, 28, 30, 93, 188, 225, 228, 230
Stressful life events 22, 127, 131, 165, 166
Substance abuse (drugs, alcohol) 7, 43, 45, 47, 183
Sudden Infant Death Syndrome (SIDS) 123, 279
Sudden loss of a loved one 3-5, 11, 26, 28, 118, 189, 278
Suicide 28, 61, 106, 127, 259
Survivor guilt 64, 98, 104
Survivor syndrome 64, 97-99

Tension 13, 19, 35, 36, 45, 53, 59, 60, 76, 80, 82, 116-118, 132, 162, 169, 171, 199, 215, 217, 219-222, 247, 253, 261, 278
Three Mile Island 69
Transference 17, 234, 235, 237, 238, 244, 273, 282
Transgenerational 102-104
Transient situational disturbances 19
Trauma inoculation 198
Trauma, definition of 15, 29, 224
Traumatic neurosis 12, 13, 16-20, 25, 30, 32, 33, 42, 46, 61
Two factor model 205-208, 247

Unemployment 22, 44

Victim support 74, 194, 196, 200, 283
Victimization 73, 151, 199
Vietnam war 32, 43-49, 163, 164, 175, 178, 283
Vietnam veterans 7, 25, 41, 43, 45, 46, 48, 171, 178, 179, 241
Violence, *see* Crimes of violence
Vulnerability 83, 100, 127, 150-153, 165, 232, 253, 277

War neurosis 32, 33, 49
Warning 2, 55, 56, 117, 162, 171, 198

Xenia tornado 66, 192

Yearning 108, 109, 111, 112, 116, 119